Lecture Notes in Mathematics

Edited by A. Dold and B. Eckmann

1254

Stephen Gelbart
Ilya Piatetski-Shapiro
Stephen Rallis

Explicit Constructions of Automorphic L-Functions

Springer-Verlag
Berlin Heidelberg New York London Paris Tokyo

Authors

Stephen Gelbart
Department of Theoretical Mathematics, The Weizmann Institute of Science
Rehovot 76100, Israel

Ilya Piatetski-Shapiro
Department of Mathematics, Tel-Aviv University
Ramat Aviv 69978, Israel
and
Department of Mathematics
Yale University
New Haven, CT 06520, USA

Stephen Rallis
Department of Mathematics, Ohio State University
Columbus, OH 43210, USA

Mathematics Subject Classification (1980): 11F70, 11S37, 22E50

ISBN 3-540-17848-1 Springer-Verlag Berlin Heidelberg New York
ISBN 0-387-17848-1 Springer-Verlag New York Berlin Heidelberg

Printed in Germany

Printing and binding: Druckhaus Beltz, Hemsbach/Bergstr.
2146/3140-543210

PREFACE

These previously unpublished manuscripts describe certain L-functions attached to automorphic representations of the classical groups.

Part A dates from 1983-84 and represents work of Piatetski-Shapiro and Rallis. The subject matter is a generalization of the method of Godement-Jacquet from $GL(n)$ to a simple classical group G.

Part B was written by Gelbart and Piatetski-Shapiro in the Fall of 1985, with an Appendix by all three authors. This work concerns a generalization of Rankin-Selberg convolution to $G \times GL(n)$, with G a classical reductive group of split rank n.

Parts A and B appear with their own Introductions and Bibliography. For a discussion of how these results are related to the recent works of F. Shahidi, the reader is referred to the 'Postscript' in the Introduction to Part B, and also to S. Gelbart and F. Shahidi's new paper on "Analytic Properties of Automorphic L-functions".

The expeditious preparation of the final form of this Lecture Note volume was done by Miriam Abraham of the Weizmann Institute, to whom we offer our thanks.

S. Gelbart

I. Piatetski-Shapiro

S. Rallis

January 1987

Research supported in part by Grant No. 84-00139 from the United States-Israel Binational Science Foundation, Jerusalem, Israel.

CONTENTS

Chapter II : The Local Functional Equation

PART A : L-FUNCTIONS FOR THE CLASSICAL GROUPS

by

I. Piatetski-Shapiro and S. Rallis

Introduction.

These notes are based on lectures given by I.I. Piatetski-Shapiro at the Institute for Advanced Study in 1983-84. The notes were prepared by J. Cogdell, who made valuable remarks improving both the mathematics and the style.

The theory described here generalizes the method of Godement-Jacquet from $GL(n)$ to the simple classical groups of symplectic, orthogonal, and unitary type. In particular, it does not require the automorphic representations whose L-functions are analyzed to be generic. Even when $G = GL(n)$ (or more generally the unit group of a matrix algebra over a division algebra) it gives a new way of looking at the Godement-Jacquet zeta-function as a Rankin-Selberg type integral involving Eisenstein series on a much larger group. The basic identity relating this Rankin-Selberg integral to a global zeta-integral for $L(\pi, s)$ is established axiomatically in §1. These axioms are then treated separately for the various different classical groups in §§2-4, and the analytaic properties of the Eisenstein series are developed in §5.

§1. A formal identity.

In this section we will present an identity, which allows us to construct L-functions with Euler product associated to irreducible cuspidal automorphic representations of certain reductive groups. The construction is based on the classic Rankin-Selberg construction. It can be applied to all of the classical groups to yield new L-functions in certain cases and new integral representations for some previously known L-functions.

We take k to be a global field, $\mathbb{A}$ the adeles of k and I_k the ideles of k. Take G to be a reductive algebraic group with anisotropic center. (This means that if C is the center of G then $C_k/C_{\mathbb{A}}$ is compact.) Let H be another reductive algebraic group over k and assume we can embed $G \times G \overset{i}{\hookrightarrow} H$. Identify $G \times G$ with its image under i and let $G^d \subset H$ be the image of G under the composition of the diagonal embedding of $G \hookrightarrow G \times G$ and the embedding $i : G \times G \hookrightarrow H$.

Let P be a parabolic subgroup of H. Then we have an action of $G \times G$ on the flag variety $X = P \setminus H$ and X will decompose into orbits under the action of $G \times G$. A $G \times G$ orbit $X' \subset X$ will be called negligible if the stabilizer R' in $G \times G$ of a point $x' \in X'$ contains the unipotent radical N' of a proper parabolic subgroup of $G \times G$ as a normal subgroup. Let $x_0 \in X$ be the point of X corresponding to the coset $P1$ and X_0 its orbit. The stabilizer R_0 of x_0 in $G \times G$ is then just $P \cap (G \times G)$.

For our construction of L-functions we need that the following two conditions on the action of $G \times G$ on X be satisfied:

(1) The stabilizer R_0 of x_0 is G^d.

(2) If X' is any orbit other than X_0, then X' is negligible.

For this reason, we will refer to X_0 as the main orbit. Justification for the term negligible will be clear from our construction.

Now assume we have H, P, and $i : G \times G \hookrightarrow H$ such that conditions (1) and (2) are satisfied. Let $\delta : P \to k^x$ be the modulus character of P and let $\omega : I_k/k^x \to \mathbb{C}$ be any quasi-character such the $\omega \circ \delta$ is trivial on $G^d_{\mathbb{A}}$. Let $f(g;\omega) \in ind_{P_{\mathbb{A}}}^{H_{\mathbb{A}}}(\omega \circ \delta)$. (Our induction ind is not normalized, i.e., $f \in ind_{P_{\mathbb{A}}}^{H_{\mathbb{A}}}(\omega \circ \delta)$ iff $f(pg) = \omega(\delta(p))f(g)$ for $g \in H_{\mathbb{A}}$

and $p \in P_{\mathbb{A}}$.) Then to f we may associate the usual Eisenstein series

$$E_f(h;\omega) = \sum_{\gamma \in P_k \backslash H_k} f(\gamma h;\omega)$$

when this is absolutely convergent. If π is an irreducible cuspidal automorphic representation of G and $\tilde{\pi}$ its contragredient, then to $\phi_1 \in \pi$ and $\phi_2 \in \tilde{\pi}$ we may associate a Rankin-Selberg type L-function by setting

$$L(\omega;\phi_1,\phi_2,f) = \int_{(G\times G)_k \backslash (G\times G)_{\mathbb{A}}} E_f((g_1,g_2);\omega)\phi_1(g_1)\phi_2(g_2)dg_1dg_2 \tag{1.1}$$

Since ϕ_1 and ϕ_2 are cuspidal, this integral converges absolutely and inherits the analytic properties of the Eisenstein series $E_f(h;\omega)$.

A key property of these L-functions is that they will have an Euler product expansion. This will follow from the following Basic Identity.

Basic Identity:

$$\int_{(G\times G)_k \backslash (G\times G)_{\mathbb{A}}} E_f((g_1,g_2);\omega)\phi_1(g_1)\phi_2(g_2)dg_1dg_2 =$$

$$= \int_{G_{\mathbb{A}}} f((g,1);\omega) < \pi(g)\phi_1,\phi_2 > dg$$

where $< \phi_1,\phi_2 >$ is the bilinear Peterson inner product given by

$$< \phi_1,\phi_2 >= \int_{G_k \backslash G_{\mathbb{A}}} \phi_1(g)\phi_2(g)dg\ .$$

Proof:

We first insert the definition of $E_f(h,\omega)$ into the integral expression (1.1) for $L(\omega;\phi_1,\phi_2,f)$ and unfold.

$$L(\omega;\phi_1,\phi_2,f) = \int_{(G\times G)_k \backslash (G\times G)_{\mathbb{A}}} \left(\sum_{P_k \backslash H_k} f(\gamma(g_1,g_2)) \right) \phi_1(g_1)\phi_2(g_2)dg_1dg_2$$

$$= \sum_{\gamma \in P_k \backslash H_k/(G\times G)_k} \int_{(G\times G)_k^{\gamma} \backslash (G\times G)_{\mathbb{A}}} f(\gamma(g_1,g_2)))\phi_1(g_1)\phi_2(g_2)dg_1dg_2$$

where $(G\times G)_k^\gamma = \{(g_1,g_2)\in (G\times G)_k | \gamma(g_1,g_2)\gamma^{-1}\in P_k\}$. Now, the double cosets $P_k\setminus H_k/(G\times G)_k$ parameterize the orbits of $(G\times G)_k$ on $X_k = P_k\setminus H_k$. We consider the main orbit and the negligible orbits separately.

a) Assume $\gamma_0\in P_k\setminus H_k/(G\times G)_k$, $\gamma_0 = 1$, corresponding to the main orbit. Then $(G\times G)_k^{\gamma_0} = (G\times G)_k\cap P_k = G_k^d$. Then

$$I(\gamma_0) = \int_{(G\times G)_k^{\gamma_0}\setminus(G\times G)_{\mathbb{A}}} f(\gamma_0(g_1,g_2);\omega)\phi_1(g_1)\phi_2(g_2)dg_1dg_2$$

$$\int_{G_k^d\setminus(G\times G)_{\mathbb{A}}} f((g_2,g_2)(g_2^{-1}g_1,1);\omega)\phi_1(g_1)\phi_2(g_2)dg_1dg_2\ .$$

Since $\omega\circ\delta$ is trivial on $G_{\mathbb{A}}^d$ we have

$$f((g_2,g_2)(g_2^{-1}g_1,1);\omega) = f((g_2^{-1}g_1,1);\omega)\ .$$

If we now write $G\times G = G^d\ \ G_1$ where $G_1 = \{(g,1)\in G\times G\}$ and write $(g_1,g_2) = (g_2,g_2)(g,1)$ with $g = g_2^{-1}g_1$ then

$$I(\gamma_0) = \int_{G_{\mathbb{A}}} f((g,1);\omega)\left(\int_{G_k\setminus G_{\mathbb{A}}}\phi_1(g_2g)\phi_2(g_2)dg_2\right)dg$$

$$= \int_{G_{\mathbb{A}}} f((g,1);\omega) < \pi(g)\phi_1,\phi_2 > dg\ .$$

b) **Negligible orbits.**

Now let $\gamma\in P_k\setminus H_k/(G\times G)_k$ correspond to a negligible orbit. If we consider the action of $G\times G$ on $P\setminus H$, the stabilizer R^γ of $P\gamma$ in $G\times G$ is

$$R^\gamma = \{(g_1,g_2)|P\gamma(g_1,g_2)\in P\gamma\} = \{(g_1,g_2)|\gamma(g_1,g_2)\gamma^{-1}\in P\} = (G\times G)^\gamma\ .$$

By the assumption that the orbit $P\gamma(G\times G)$ is negligible, there is a proper parabolic $P^\gamma\subset G\times G$ whose unipotent radical N^γ is normal in R^γ. Then

$$\begin{aligned}I(\gamma) &= \int_{R_k^\gamma\setminus(G\times G)_{\mathbb{A}}} f(\gamma(g_1,g_2);\omega)\phi_1(g_1)\phi_2(g_2)dg_1dg_2\\ &= \int_{R_{\mathbb{A}}^\gamma\setminus(G\times G)_{\mathbb{A}}}\Big(\int_{R_k^\gamma\setminus R_{\mathbb{A}}^\gamma} f(\gamma(r_1,r_2)(g_1',g_2'))\phi_1(r_1g_1')\phi_2(r_2g_2')dr_1dr_2\Big)dg_1'dg_2'\ .\end{aligned}$$

If in the inner sum we integrate first over $N_k^\gamma \setminus N_{\mathbb{A}}^\gamma$, the result is a function on $M_k \setminus M_{\mathbb{A}}$ where $M = N^\gamma \setminus R^\gamma$. Hence we may write

$$\int_{R_k^\gamma \setminus R_{\mathbb{A}}^\gamma} f(\gamma(r_1,r_2)(g_1',g_2'))\phi_2(r_1g_1')\phi_2(r_2g_2')dr_1dr_2 =$$

$$= \int_{M_k\setminus M_{\mathbb{A}}} \left(\int_{N_k^\gamma \setminus N_{\mathbb{A}}^\gamma} f(\gamma(n_1,n_2)(m_1,m_2)(g_1',g_2')) \right.$$
$$\left. \phi_1(n_1m_1g_1')\phi_2(n_2m_2g_2') \cdot dn_1dn_2 \right) dm_1dm_2 \; .$$

If we now write $N^\gamma = N_1 \times N_2$ with N_i the unipotent radical of some parabolic $P_i \subset G$ (at least one non-trivial), then, since δ is trivial on the unipotent elements of $P_{\mathbb{A}}$, we have the above integral equal to

$$\int_{M_k\setminus M_{\mathbb{A}}} f(\gamma(m_1,m_2)(g_1',g_2')) \left(\int_{N_{1,k}\setminus N_{1,\mathbb{A}}} \phi_1(n_1m_1g_1')dn_1 \right)$$
$$\left(\int_{N_{2,k}\setminus N_{2,\mathbb{A}}} \phi_2(n_2m_2g_2')dn_2 \right) \cdot dm_1dm_2 \; .$$

Since ϕ_1 and ϕ_2 are cusp forms, at least one of the integrals

$$\int_{N_{i,k}\setminus N_{i,\mathbb{A}}} \phi_i(n_im_ig_i')dn_i$$

is identically zero. This implies that for the negligible orbits $I(\gamma) \equiv 0$ and hence they contribute nothing to $L(\omega;\phi_1,\phi_2,f)$ (thus justifying the term negligible). This completes the proof.

Due to the uniqueness of the pairing of π with $\tilde{\pi}$, the global Peterson bilinear form $<,>$ decomposes into a product of local invariant forms $<,>_v$ in the sense that if $\phi_1 = \Pi_v\phi_{1,v} \in \pi$ and $\phi_2 = \Pi_v\phi_{2,v} \in \tilde{\pi}$ are decomposable, then $<\phi_1,\phi_2> = \Pi_v <\phi_{1,v},\phi_{2,v}>_v$. Keeping the basic identity in mind, we now define the local version of our L-functions. For $f \in ind_{P_v}^{H_v}(\omega_v \circ \delta_v), \phi_{1,v} \in \pi_v$, and $\phi_{2,v} \in \tilde{\pi}_v$, define

$$L_v(\omega_v;\phi_{1,v},\phi_{2,v},f_v) = \int_{G_v} f((g,1);\omega_v) < \pi_v(g)\phi_{1,v},\phi_{2,v} >_v dg_v \; .$$

Then as a corollary to the basic identity we have the following.

Corollary. The global L-function admits an Euler product given by

$$L(\omega,\phi_1,\phi_2,f) = \Pi_v L_v(\omega_v;\phi_{1,v},\phi_{2,v},f_v)$$

for $\phi_1 \in \pi, \phi_2 \in \tilde{\pi}$, and $f \in ind_{P_{\mathbb{A}}}^{H_{\mathbb{A}}}(\omega \circ \delta)$ all decomposable.

Remark: A priori, one might try to define a global L-function as in (1.1) for $\phi_i \in \pi_i$ with π_i arbitrary irreducible cuspidal automorphic representations. But, by the Basic Identity, these would all be identically zero unless $\pi_2 \simeq \tilde{\pi}_1$ since $< \pi(g)\phi_1, \phi_2 > \equiv 0$ unless $\phi_2 \in \tilde{\pi}_1$.

§2. Explicit constructions for the symplectic, orthogonal, and unitary groups.

In this section we will restrict our attention to the classical groups G of symplectic, orthogonal, or unitary type. In these cases we will explicitly construct a group H, an embedding $i : G \times G \hookrightarrow H$, and a parabolic P of H satisfying the conditions (1) and (2) on the orbits of $G \times G$ in $X = P \setminus H$.

We begin by setting up some common notation. Let k be a global field.

(i) Symplectic groups: Let V be a vector space of even dimension $n = 2m$ over k and let $(\, , \,)$ be a non-degenerate skew-symmetric form on V. Let $G \subset GL(V)$ be the isometries of this form. Then $G = Sp(n)$.

(ii) Orthogonal groups: Let V be a vector space of (arbitrary) dimension n over k and let $(\, , \,)$ denote a non-degenerate symmetric bilinear form on V. Let $G \subset GL(V)$ be the group of isometries of $(\, , \,)$, so $G = O(n)$.

(iii) Unitary groups: Let K' be a quadratic extension of k. Let V be a vector space over K' of dimension n and let $(\, , \,)$ be a non-degenerate Hermitian form on V with respect to the non-trivial automorphism of K' over k. Then let $G \subset GL(n, K')$ be the group of transformations preserving $(\, , \,)$, so that $G = U(n)$.

Now take G, V and $(\, , \,)$ to be as in any of the cases (i), (ii), or (iii). In cases (i) and (ii) set $K = k$ and in case (iii) set $K = K'$. We will construct H by "doubling the variables" . Let $W = V \oplus V$ and define a form $<,>$ on W by

$$< (v_1, v_2), (v_1', v_2') >= (v_1, v_1') - (v_2, v_2').$$

Then $<,>$ is non-degenerate and of the same type as $(\, , \,)$. The form $<,>$ admits isotropic subspaces of maximal dimension. In fact, if we let $V^d = \{(v, v) \in W\}$ be the image of the diagonal embedding of V in W then $dim_K(V^d) = n = \frac{1}{2} dim_K(W)$ and V^d is isotropic for $<,>$.

Now let $H \subset GL(2n, K)$ be the group of isometries of $<,>$, and let P be the parabolic subgroup of H preserving V^d . There is an embedding $i : G \times G \hookrightarrow H$ by letting

$(v_1, v_2) \cdot i(g_1, g_2) = (v_1 g_1, v_2 g_2)$ for $v_1, v_2 \in V$ and $g_1, g_2 \in G$. Identify $G \times G$ with its image under i .

To show that H, P , and the embedding $i : G \times G \hookrightarrow H$ satisfy conditions (1) and (2) of Section 1 we must investigate the orbit structure of $X = P \setminus H$ under $G \times G$. Let $i_+ : V \hookrightarrow W$ be given by $i_+(v) = (v, o)$ and $i_- : V \hookrightarrow W$ be given by $i_-(v) = (o, v)$. Let $V^\pm$ be the image of $i_\pm$. Let L be any maximal isotropic subspace of W. Then let $L^+ = L \cap V^+$, $L^- = L \cap V^-, \kappa^+(L) = dim_K(L^+)$, and $\kappa^-(L) = dim_K(L^-)$. Since H acts transitively on the space of maximal isotropic subspaces of W and P stabilizes V^d, then we may view $X = P \setminus H$ as the variety of maximal isotropic subspaces of W .

Lemma 2.1.: Let L be a maximal isotropic subspace of W . Then $\kappa^+(L) = \kappa^-(L) = \kappa(L)$ and $\kappa(L)$ is the only invariant of the $G \times G$ orbit of L in X ; in other words, if $\kappa(L) = \kappa(M)$ for $L, M \in X$, then there exists $g \in G \times G$ such that $Lg = M$.

Proof: Let $\pi^\pm$ be the orthogonal projection of W onto $V^\pm$. Let $L' = \pi^+(L)$ and $L'' = \pi^-(L)$. Since $L^\mp$is the kernel of $\pi^\pm$ restricted to L, $dim_K L = dim_K L' + dim_K L^- = dim_K L'' + dim_K L^+$. On the other hand, $L^+ \subset L'$ and L^+ is in the kernel of the form $(\, , \,)$ restricted to L' . Since the form is non-degenerate on V^+ there must be a subspace $L_+ \subset V^+$, of the same dimensions as L^+ , which pairs non-degenerately with L^+ . Therefore $L' \oplus L_+ \subset V^+$, so that $dim_K L' + dim_K L^+ \leq dim_K V = dim_K L' + dim_K L^-$, and hence $dim_K L^+ \leq dim_K L^-$. Similarly $dim_K L^- \leq dim_K L^+$, so that $\kappa^+(L) = dim_K L^+ = dim_K L^- = \kappa^-(L)$. Note that this implies that L' is the full orthogonal subspace to L^+ in V^+ and similarly for L'' and L^- in V^- .

Since L^+ is the kernel of the form $(\, , \,)$ restricted to L' , $(\, , \,)$ induces a non-degenerate form on $L^+ \setminus L'$. Similarly for $L^- \subset L''$. Let $\pi_1 : L' \to L^+ \setminus L'$ and $\pi_2 : L'' \to L^- \setminus L''$ be the projections. Then the isotropic subspace L defines a isometry $g_L : L^+ \setminus L' \to L^- \setminus L''$ by $(\pi_1 v_1) g_L = (\pi_2 v_2)$ iff $(v_1, v_2) \in L$. This is seen to be well-defined and is an isometry since L is totally isotropic in W. Furthermore, the spaces L', L^+, L'', L^- and the isometry g_L completely determine L, for $L = \{(v_1, v_2) : v_1 \in L', v_2 \in L''$ and $(\pi_1 v_1) g_L = \pi_2 v_2\}$.

That $\kappa(L)$ is an invariant of the $G \times G$ orbit of L in X is evident, since $(L(g_1, g_2))^+ = L^+ g_1$. Now we will show that if L and M are totally isotropic subspaces of W with $\kappa(L) = \kappa(M)$ then there exists $g = (g_1, g_2) \in G \times G$ such that $Lg = M$. Since L^+ and M^+ are isotropic in V^+ of the same dimension, there is $L^+ g_1 = M^+$. Similarly there is

$g_2 \in G$ such that $L^- g_2 = M^-$. So replacing L by $L(g_1, g_2)$ we may assume $L^+ = M^+$ and $L^- = M^-$. Then since L' is the orthocomplement of L^+ , and the same is true for M' , we have $L' = M'$. Similarly $L'' = M''$. Then L and M both define isometries $g_L, g_M : L^+ \setminus L' \to L^- \setminus L''$. These will differ by an isometry γ of $L^- \setminus L''$, i.e., $g_L = g_M \gamma$ with $\gamma : L^- \setminus L'' \to L^- \setminus L''$. γ may be lifted to an isometry γ'' of L'' satisfying $\pi_2(v_2\gamma'') = (\pi_2 v_2)\gamma$ and this may be extended, via Witt's theorem, to an isometry γ'' of V^-. We claim that $L(1,\gamma'')^{-1} = M$, with $(1,\gamma'') \in G \times G$. We have $(v_1, v_2) \in L$ iff $(\pi_1 v_1) g_L = (\pi_2 v_2)$. Therefore $L(1,\gamma'')^{-1} = \{(v_1, v_2) \in L' \times L'' : (\pi_1 v_1) g_L = \pi_2(v_2\gamma'')\}$. But $g_L \gamma^{-1} = g_M$. Therefore $(v_1, v_2) \in L(1,\gamma'')^{-1}$ iff $(\pi_1 v_1) g_M = (\pi_2 v_2)$ iff $(v_1, v_2) \in M$. This completes the proof.

Now for $0 \leq d \leq n$, let X_d be the $G \times G$ orbit in X of maximal isotropic subspaces L with $\kappa(L) = d$. Then since $\kappa(V^d) = 0$ we have V^d in the orbit X_0 . The stabilizer of V^d in $G \times G$ is $(G \times G) \cap P$. On the other hand, an element $(g_1, g_2) \in G \times G$ stabilizes V^d iff $vg_1 = vg_2$ for all $v \in V$, i.e., iff $g_1 = g_2$. So indeed $G^d = (G \times G) \cap P$ and condition (1) is satisfied.

To show that condition (2) is satisfied, we must show that if $d > 0$ then the orbit X_d is negligible. So fix $d > 0$ and let $L \in X_d$. Let P^+ be the parabolic subgroup of G preserving the flag $V \supset L' \supset L^+$ and P^- the parabolic subgroup of G preserving the flag $V \supset L'' \supset L^-$. Since $d > 0$ these are proper parabolics. Let $N^\pm$ be the unipotent radical of $P^\pm$ so that $N = N^+ \times N^-$ is then the unipotent radical of the proper parabolic $P^+ \times P^-$ of $G \times G$. Now let R be the stabilizer of L in $G \times G$. R can be characterized as the pairs $(g_1, g_2) \in P^+ \times P^-$ such that $(\pi_1(v_1 g_1)) g_L = \pi_2(v_2 g_2)$. (Recall that $\pi_1 : L' \to L^+ \setminus L'$ and $\pi_2 : L'' \to L^- \setminus L''$ are the canonical projections.) Since N is normal in $P^+ \times P^-$ we need only show that $N \subset R$. But by definition, N^+ induces the identity on $L^+ \setminus L'$ and N^- induces the identity on $L^- \setminus L''$. So for $(v_1, v_2) \in L' \times L''$ and $(n_1, n_2) \in N$ we have $\pi_1(v_1 n_1) = \pi_1(v_1)$ and $\pi_2(v_2 n_2) = \pi_2(v_2)$. So $(\pi_1(v_1 n_1)) g_L = (\pi_1(v_1)) g_L = \pi_2(v_2) = \pi_2(v_2 n_2)$. Then $N \subset R$. This shows that for $d > 0$ the orbit X_d is negligible. Therefore we have proved the following proposition.

Proposition 2.1: The choices of group H, parabolic P and embedding $i : G \times G \to H$ above satisfy conditions (1) and (2) of Section 1.

§3. Explicit constructions for $PGL(n)$. Connections with the work of Godement and Jacquet.

Let D be a central simple division algebra of degree m over k and let $G = PGL(n, D)$. In this section we will construct a group H, parabolic subgroup $P \subset H$ and an embedding $i : G \times G \hookrightarrow H$ satisfying conditions (1) and (2) of Sect. 1. In this situation, the L-functions of Section 1 will be the same as the zeta functions considered by Godement and Jacquet in [G-J].

3.1. Let $V = M(n, D)$. As a vector space over k, V has dimension $N = n^2m^2$. There is a natural action of $GL(n, D) \times GL(n, D)$ on V by $x \cdot (g_1, g_2) = g_2^{-1}xg_1$ for $g_i \in GL(n, D)$ and $x \in M(n, D)$. This gives a homomorphism $GL(n, D) \times GL(n, D) \to GL(N, k)$ which will induce an embedding $i : G \times G \hookrightarrow PGL(N, k)$. We will take $H = PGL(N, k)$ and $i : G \times G \hookrightarrow H$ this embedding. Identify $G \times G$ with its image. Let $e_0 \in V$ correspond to the identity $1_n \in M(n, D)$. Take P to be the parabolic subgroup of H stabilizing the k-line through e_0 .

To show that conditions (1) and (2) of Sect. 1 are satisfied, we must determine the orbit structure of $X = P \setminus H$ under the action of $G \times G$. Since P stabilizes a k-line in V, X is the variety of k-lines in V . For $x \in V$ we will let $< x >$ denote the k-line spanned by x. Let $W = D^n$ be a vector space over D considered as a space of row vectors. W is a D-module under left multiplication and a $M(n, D)$ module under right matrix multiplication. As a vector space over k, $dim_k(W) = nm^2$. For each $x \in V$ we may define an invariant of the $G \times G$ orbit of $< x >$ in X as follows. Viewing x as an element of $M(n, D)$, Wx will be a subspace of W. If $y = \lambda x$ with $\lambda \in k^v$ then $Wy = Wx$ and hence Wx depends only on the span $< x >$ of x. Then define $\kappa(x) = dim_k(Wx)$.

<u>**Lemma 3.1**</u>: For $x \in V$, $\kappa(x)$ is the only invariant of the $G \times G$ orbit of $< x >$ in X, in other words if $\kappa(x_1) = \kappa(x_2)$, $x_1, x_2 \in V$ then there exists a $g \in G \times G$, such that $< x_1 > g = < x_2 >$.

<u>**Proof:**</u> We have already seen that $\kappa(x)$ depends only on the span $< x >$ of x. To show that $\kappa(x)$ is an invariant of the $G \times G$ orbit it will suffice to show that for $g_1, g_2 \in GL(n, D)$, $\kappa(g_2^{-1}xg_1) = \kappa(x)$. But this is clear since $\kappa(g_2^{-1}xg_1) = dim_k(Wg_2^{-1}xg_1) =$

$dim_k(Wxg_1) = dim_k(Wx)$, using that g_1 and g_2 are invertible. So $\kappa(x)$ is indeed an orbit invariant.

To see that $\kappa(x)$ is the only orbit invariant, we must show that if $x, y \in M(n, D)$ are such that $\kappa(x) = \kappa(y)$ then there are $g_1, g_2 \in GL(n, D)$ such that $< x > = < g_2^{-1} y g_1 >$. Since $\kappa(x) = \kappa(y)$ we know that $dim_k(Wx) = dim_k(Wy)$. Both Wx and Wy are D-linear subspaces of W (D acting on the left) of the same dimension, so there is an element $g_1 \in GL(n, D)$ such that $Wx = Wyg_1$. Then since both x and yg_1 are D-linear maps from $W \to Wx = Wyg_1$ they differ by a D-automorphism of W , i.e., there is some $g \in GL(n, D)$ such that $x = gyg_1$. Taking $g_2 = g^{-1}$ we have $x = g_2^{-1} y g_1$ and so, of course, $< x > = < g_2^{-1} y g_1 >$. This proves the lemma.

The orbit invariant $\kappa(x)$ takes the values $0, m^2, 2m^2, \cdots, nm^2$. So for $0 \leq d \leq n$ let X_d be the $G \times G$ orbit in X given by $X_d = \{< x > \in X : \kappa(x) = (n - d)m^2\}$. Then X_0 is the orbit through the line $< e_0 >$ spanned by the identity element of $M(n, D)$. Let $x_0 = e_0$ and for $d = 1, \cdots, n$ let $x_d \in V$ such that $< x_d > \in X_d$. Let R_d be the stabilizer in $G \times G$ of $< x_d >$. So $R_0 = P \cap (G \times G)$.

Proposition 3.1. The main orbit X_0 satisfies $R_0 = P \cap G \times G = G^d$. The orbits X_d with $d \geq 1$ are negligible. So this choice of H, P, and $i : G \times G \to H$ satisfies conditions (1) and (2) of Sect. 1.

Proof: First consider the main orbit X_0 . Let $(g_1, g_2) \in G \times G$ and let $\tilde{g}_1, \tilde{g}_2 \in GL(n, D)$ be any elements projecting to g_1 and g_2. Then (g_1, g_2) will stabilize $< x_0 >$ iff $< \tilde{g}_2^{-1} \tilde{g}_1 > = < x_0 >$ iff $\tilde{g}_1 = \lambda \tilde{g}_2$ with $\lambda \in k^x$ iff $g_1 = g_2$. Therefore $R_0 = G^d$.

Next let $d \geq 1$ and let $L_d = Wx_d$. Let $Q_d = \{g \in PGL(n, D) : L_d g = L_d\}$ be the parabolic subgroup of G stabilizing L_d . Since $dim_k(L_d) = (n - d)m^2 < nm^2$, Q_d is a proper parabolic. Let N_d be its unipotent radical. Now, $R_d = \{(g_1, g_2) \in G \times G : \tilde{g}_2^{-1} x \tilde{g}_1 = \lambda x_d$ with $\lambda \in k^\times\}$. So if $(g_1, g_2) \in R_d$, $L_d g_1 = L_d \tilde{g}_1 = W x_d \tilde{g}_1 = W \tilde{g}_2^{-1} x_d \tilde{g}_1 = W \lambda x_d = L_d$. Therefore $g_1 \in Q_d$. Since N_d is normal in Q_d , the group $N = N_d \times 1$ will be normal in R_d . Since N is the unipotent radical of the proper parabolic subgroup $Q_d \times G$ of $G \times G$, this shows that X_d is negligible for $d \geq 1$ and finishes the proposition.

3.2. Using the choice of H, P, and $i : G \times G \to H$ made here, the construction of Sect. 1 gives a family of L-functions with Euler products attached to an automorphic cuspidal representation π of G . On the other hand, if we consider π as a representation of $GL(n, D)$ having trivial central character, Godement and Jacquet $[G-J]$ have assigned a family of zeta functions with Euler products to π . We would now like to compare these two constructions.

We first recall the Godement-Jacquet construction. With π as above, let $\phi_1 \in \pi$ and $\phi_2 \in \tilde{\pi}$, the contragredient representation. Let $\xi(g)$ be the matrix element $\xi(g) = < \pi(g)\phi_1, \phi_2 >$. Then for $\Phi \in S(M(n, D_{\mathbb{A}}))$ a Schwartz-Bruhat function, Godement and Jacquet define a zeta function $Z(\Phi, s, \xi)$ by

$$Z(\Phi, s, \xi) = \int_{GL(n, D_{\mathbb{A}})} |\nu(det\ g)|^s \Phi(e_0 g)\xi(g)dg$$

where $\nu : D \to k$ is the reduced norm.

We will show how to obtain these functions out of our Rankin-Selberg construction. With$\Phi \in S(M(n, D_{\mathbb{A}}))$ as above and $h \in GL(N, K)$, let

$$f^{\Phi}(h; s) = |det\ h|^s \int_{I_k} \Phi(ae_0 h)|a|^{NS} d^x a\ .$$

(Recall that $N = m^2 n^2 = dim_k(M(n, D))$.) Since f^{Φ} is invariant under the center of $GL(N, k)$ we may consider f^{Φ} as a function on H . Then f^{Φ} is in some induced representation from $P_{\mathbb{A}}$. Let $E^{\Phi}(h; S) = \sum_{\gamma \in P_k \backslash H_k} f^{\Phi}(\gamma h; s)$ be the associated Eisenstein series. Then if $\phi_1 \in \pi$ and $\phi_2 \in \tilde{\pi}$ we may form the L-function

$$L(s; \phi_1, \phi_2, f^{\Phi}) = \int_{(G \times G)_k \backslash (G \times G)_{\mathbb{A}}} E^{\Phi}((g_1, g_2); s)\phi_1(g_1)\phi_1(g_2)dg_1 dg_2$$

as in Section 1.

As a direct consequence of the Basic Identity of Section 1 we have the following result.

Proposition 3.2: With the notation as above,

$$L(s; \phi_1, \phi_2, f^{\Phi}) = Z(\Phi, ns, \xi)$$

with $\xi(g) = < \pi(g)\phi_1, \phi_2 >$.

Proof: Applying the Basic Identity we have

$$L(s; \phi_1, \phi_2, f^{\Phi}) = \int_{G_{\mathbb{A}}} f^{\Phi}((g, 1); s) < \pi(g)\phi_1, \phi_2 > dg\ .$$

Substituting the definition of f^{Φ} and ξ we have

$$\begin{aligned} L(s;\phi_1,\phi_2,f^{\Phi}) &= \int_{G_{\mathbb{A}}} |det(g,1)|^s \Big(\int_{I_k} \Phi(ae_0(g,1))|a|^{Ns} d^x a\Big)\xi(g)dg \\ &= \int_{G_{\mathbb{A}}} |\nu(det\ g)|^{ns} \Big(\int_{I_k} \Phi(e_0 ag)|a|^{Ns} d^x a\Big)\xi(g)dg \\ &= \int_{GL(n,D_{\mathbb{A}})} |\nu(det\ g)|^{ns}\Phi(g)\xi(g)dg \\ &= Z(\Phi, ns, \xi)\ . \end{aligned}$$

§4. Remarks on $GL(n)$ and the other classical groups.

The method of doubling variables in Section 2 also gives explicit constructions for the groups H, P , and embedding $i : G \times G \to H$ for the other classical groups as well. In addition, if we modify slightly the construction of the L-functions given in Sect. 1 we can use the doubling variables technique to construct L-functions for $GL(n)$, even though the center is not anisotropic. In this section we will indicate how these are done but will omit the details of the proofs since they are essentially as in Sect. 1 and 2.

4.1. Let us begin with the other classical groups with anisotropic center. They fall into two categories, which we now describe.

First there are the quaternionic Hermitian groups. Keep k a global field as before and let D be a division quaternion algebra over k. Let $x \to x^\sigma$ be the main involution of D over k , so $x + x^\sigma \in k$ and $xx^\sigma \in k$ for $x \in D$. Let $\mathfrak{R} = M(n, D)$. Extend σ to $\mathfrak{R}$ by setting $(a_{ij})^\sigma = (b_{ij})$ with $b_{ij} = a_{ji}^\sigma$. Then let $S \in \mathfrak{R}^x$ satisfying $S^\sigma = \varepsilon S$ with $\varepsilon = \pm 1$. Let G be the quaternionic Hermitian group defined by $G = \{g \in \mathfrak{R}^x : gSg^\sigma = S\}$.

When $\varepsilon = 1$, G is a form of a symplectic group and when $\varepsilon = -1$, G is a form of an orthogonal group. To see this, let $\bar{k}$ be the algebraic closure of k . Then $D \otimes_k \bar{k} = M(2, \bar{k})$ and $\mathfrak{R} \otimes_k \bar{k} = M(2n, \bar{k})$. The main involution σ of D over k extends to an involution of $M(2, \bar{k})$ over $\bar{k}$. By the Skolem-Noether Theorem, any involution on $M(2, \bar{k})$ is equivalent to transposition by an inner automorphism. Hence there exists $\alpha \in GL(2, \bar{k})$ such that for $\beta \in M(2, \bar{k})$, $\beta^\sigma = \alpha {}^t\beta \alpha^{-1}$. Then since $(\beta^\sigma)^\sigma = \beta$, α must satisfy ${}^t\alpha = \pm\alpha$. On the other hand, since σ satisfies $dim_k\{\beta \in M(2, \bar{k}) : \beta^\sigma = \beta\} = 1$, α must satisfy ${}^t\alpha = -\alpha$. So α is skew symmetric. Now write $M(2n, \bar{k})$ in 2×2 block form (corresponding to the 1×1 block form of $M(n, D)$). Then if A is the block diagonal matrix with α in the diagonal positions, the extension of σ to $M(2n, \bar{k})$ is given by $B^\sigma = A^tBA^{-1}$ for $B \in M(2n, \bar{k})$. Then $G(\bar{k}) = \{g \in M(2n, \bar{k}) | gSg^\sigma = S\} = \{g \in M(2n, \bar{k}) | gSA^tg = SA\}$ and ${}^tA^tS = (-A)(A^{-1}S^\sigma A)) = -\varepsilon(SA)$. So the matrix SA defining $G(\bar{k})$ is symmetric for $\varepsilon = -1$ and skew-symmetric for $\varepsilon = 1$.

Next let K be a quadratic extension of k and let D be a division algebra of degree m over K which admits an involution σ of the second kind, i.e., an involution $\sigma : D \to D$

which induces the nontrivial Galois automorphism of K over k when restricted to K. Again, let $\mathfrak{R} = M(n, D)$ and let $S \in \mathfrak{R}^x$ satisfy $S^\sigma = \varepsilon S$ with $\varepsilon = \pm 1$. In this case we take

$$G = \{g \in \mathfrak{R} | gSg^\sigma = S\} .$$

G is then a form of $GL(nm)$.

In either of these cases we let $V = D^n$ as a space of row vectors. Then G acts on V on the right. Let $W = V \oplus V$. If we set

$$\tilde{S} = \begin{pmatrix} S & 0 \\ 0 & -S \end{pmatrix} \quad \in M(2n, D)$$

then $\tilde{S}$ defines a D-valued σ-Hermitian form on W and we take

$$H = \{g \in M(2n, D) : g\tilde{S}g^\sigma = \tilde{S}\} .$$

Then $G \times G \hookrightarrow H$ naturally (along the diagonal in block form) and we take this embedding to be i . If we let $V^d \subset W$ be the subspace $\{(v, v) \in W = V \oplus V\}$ and let P be the parabolic subgroup preserving V^d, then H, P, and $i : G \times G \to H$ satisfy the conditions (1) and (2) of Sect. 1. The arguments are essentially those of Sect. 2.

Notice that with respect to the σ-Hermitian form $<,>$ on W defined by $\tilde{S}$, the subspace V^d is an isotropic subspace of maximal dimension. Since this form is nondegenerate, there is an isotropic complement V_d to V^d such that $W = V^d \oplus V_d$ and $<,>$ allows the identification of V_d with the D-linear dual of V^d . Choosing a basis of V^d and the dual basis for V_d we obtain a basis for W in which the matrix $\tilde{S}'$ representing $<,>$ takes the form

$$\tilde{S}' = \begin{cases} \begin{pmatrix} & & 1_D \\ & \cdot^{\cdot^{\cdot}} & \\ 1_D & & \end{pmatrix} & if \ \ \varepsilon = -1 \\[2ex] \begin{pmatrix} & & & & & 1_D \\ & & & & \cdot^{\cdot^{\cdot}} & \\ & & & 1_D & & \\ & & -1_D & & & \\ & \cdot^{\cdot^{\cdot}} & & & & \\ -1_D & & & & & \end{pmatrix} & if \ \ \varepsilon = +1 . \end{cases}$$

Therefore the group H used in the construction depends only upon D, n, and ε and not upon the particular choice of S defining G. Of course, the choice of P and $i : G \times G \to H$ reflects this choice of S.

4.2. Now take D to be any central simple division algebra over k and let $G = GL(n, D)$. We will first use the doubling of variables technique to describe H, P, and $i : G \times G \to H$ which satisfy the conditions (1) and (2) of Sect. 1. Then we will show how to modify the L-function construction to deal with the fact that the center of G is not anisotropic.

Let $V = D^n$ be a vector space over D (and hence over k) considered as a space of row vectors. Then D^x acts on the left by scalar multiplication and G acts on the right in the usual manner. Let $W = V \oplus V$ and let $H = GL(2n, D)$ realized as the k-linear automorphisms of W (acting on the right) which commute with the action of D^x on the left. Then $i : G \times G \to H$ is the natural embedding along the diagonal. As the parabolic P we take the parabolic of H which stabilizes the subspace

$$V^d = \{(v, v) \in W = V \oplus V : v \in V\} \quad \text{of} \quad W .$$

Then H, P, and $i : G \times G \to H$ satisfy conditions (1) and (2) of Sect. 1.
Characterization of the orbit structure of $X = P \setminus H$ is as follows. X can be realized as the variety of k-subspaces of W which are stable under the D^x action on the left and are of dimension nm^2 over k (where m is the degree of D/k). Let V^+ be the subspace $V^+ = \{(v, 0) \in W : v \in V\}$ and $V^- = \{(0, v) \in W : v \in V\}$.

Lemma 4.1. Let L be a subspace in X. Let $\kappa^+(L) = dim_k(L \cap V^+)$ and $\kappa^-(L) = dim_k(L \cap V^-)$. Then the pair $(\kappa^+(L), \kappa^-(L))$ is the only invariant of the $G \times G$ orbit of L in X.

The proof of this lemma and of the fact that H, P, and i satisfy conditions (1) and (2) are essentially those of Sect. 2.

To describe the L-functions in this setting, begin with $f(h; \omega) \in ind_{P_{\mathbb{A}}}^{H_{\mathbb{A}}}(\omega \circ \delta)$ and $E_f(h; \omega)$ just as in Sect. 1. Then take $\alpha \in C_c^\infty(\mathbb{R}_+^x)$. If π is an irreducible cuspidal automorphic representation of G, $\tilde{\pi}$ its contragredient, $\phi_1 \in \pi$, and $\phi_2 \in \tilde{\pi}$, we define

$$L(\omega; \phi_1, \phi_2, f, \alpha) =$$

$$\int_{C_{\mathbb{A}}(G\times G)_k\backslash(G\times G)_{\mathbb{A}}} E_f((g_1,g_2);\omega)\phi_1(g_1)\phi_1(g_2)\alpha(|det(\frac{g_1}{g_2})|)dg_1dg_2$$

where now C is the center of H. Due to the presence of α, this integral is convergent. In this framework the Basic Identity becomes

$$L(\omega;\phi_1,\phi_2,f,\alpha) = \int_{G_{\mathbb{A}}} f((g,1);\omega) < \pi(g)\phi_1,\phi_2 > \alpha(|det\ g|)dg \tag{4.1.}$$

where now $< \phi_1,\phi_2 >= \int_{G_k C_{\mathbb{A}}\backslash G_{\mathbb{A}}} \phi_1(g)\phi_2(g)dg$. The integral on the right hand side of (4.1) is absolutely convergent in some halfplane independent of $\alpha \in C_c^{\infty}(\mathbb{R}_+^x)$ and in fact is absolutely convergent for $\alpha \equiv 1$. If we then take $\{\alpha_n\}$ a sequence in $C_c^{\infty}(\mathbb{R}_n^x)$ which is monotonic increasing and $\lim_{n\to\infty} \alpha_n = 1$ then we may define

$$\begin{aligned} L(\omega;\phi_1,\phi_2,f) &= \lim_{n\to\infty} L(\omega;\phi_1,\phi_2,f,\alpha_n) \\ &= \int_{G_{\mathbb{A}}} f((g,1);\omega) < \pi(g)\phi_1,\phi_2 > dg. \end{aligned}$$

This is independent of the choice of the sequence $\{\alpha_n\}$, as the last equality shows, and extends the construction of Sect. 1 to $G = GL(n,D)$.

4.3. It is possible to modify the conditions (1) and (2) in order to cover some cases where, instead of an embedding, there exists only a homomorphism $G \times G \to H$. In this way our construction will yield L-functions for G a spinor group or a Metaplectic group. However, we will not pursue this matter any further here.

§5. Special Eisenstein Series for $O(2n)$ and $Sp(2n)$.

The strength of the Rankin-Selberg method of constructing L-functions is that the analytic properties of the L-function are obtained from those of the Eisenstein series. For a general Eisenstein series the desired analytic properties are not always available. However, for the construction of our L-functions a relatively simple family of Eisenstein series can be used. These Eisenstein series will have a meromorphic continuation to the whole complex plane with only a finite number of poles, whose possible location can be determined, and will satisfy a simple functional equation. In this section we describe these Eisenstein series for the groups $Sp(2n)$ and the split $O(2n)$.

5.1. Before we define the Eisenstein series let us recall some facts about the structure of $Sp(2n)$ and $O(2n)$ which will be necessary in the sequel. Fix k a global field. Let $G = Sp(n)$ or $O(n)$. In Section 2 we have constructed from G a group $H = Sp(2n)$ or $O(2n)$ respectively which is split over k , i.e., such that the standard representation of H admits isotropic subspaces of maximal dimension. We will retain the notation of Section 2. So G acts on the vector space V, H acts on $W = V \oplus V$, and P is the maximal parabolic subgroup of H preserving the maximal isotropic subspace V^d .

Since the form $<,>$ is non-degenerate on W , there is a complimentary isotropic subspace V_d to V^d so that $W = V_d \oplus V^d$ and $<,>$ gives a non-degenerate pairing of V_d and V^d . Take a basis of W consisting of a basis of V^d and the dual basis for V_d . If we then set

$$S_H = \begin{pmatrix} 0 & 1_n \\ \varepsilon_H 1_n & 0 \end{pmatrix}$$

with

$$\varepsilon_H = \begin{cases} 1 & \mathrm{H} = \mathrm{O(2n)} \\ -1 & \mathrm{H} = \mathrm{Sp(2n)} \end{cases}$$

then $H_k = \{h \in GL(2n,k) | hS^t_H h = S_H\}$. The parabolic P then has the Levi decomposition $P = MU$ with

$$M_k = \left\{ \begin{pmatrix} g & 0 \\ 0 & {}^tg^{-1} \end{pmatrix} \mid g \in GL(n,k) \right\}$$

$$U_k = \left\{ \begin{pmatrix} 1 & u \\ 0 & 1 \end{pmatrix} \mid u \in M(n,k), \ \ {}^tu + \varepsilon_H u = 0 \right\} .$$

As a maximal k-split torus $T \subset P$ we take

$$T_k = \left[\underline{t} = \begin{pmatrix} t_1 & & & & & \\ & \ddots & & & & \\ & & t_n & & & \\ & & & t_1^{-1} & & \\ & & & & \ddots & \\ & & & & & t_n^{-1} \end{pmatrix} \mid t_i \in k^x \right] .$$

Let B be the standard Borel subgroup of H with $P \supset B \supset T$ and let N be the unipotent radical of B .

Let $X^*(T)$ be the group of rational characters of T. The characters $x_i : t \mapsto t_i$ from a $\mathbb{Z}$-basis of $X^*(T)$. Let $\Phi_H = \Phi(H,T)$ be the set of roots of T in H, Φ_H^+ the set of positive roots of Φ_H determined by N. Since T is also a maximal split torus in M we have $\Phi_M = \Phi(M,T)$ the set of roots of T in M and $\Phi_M^+ = \Phi_M \cap \Phi_H^+$. In the basis $\{x_1 \cdots x_n\}$ of $X^*(T)$ we have:

$$\Phi_H = \begin{cases} \{\pm(x_i \pm x_j) | i < j\} & \text{H=0(2n)} \\ \{\pm(x_i \pm x_j), \pm 2x_i | i < j & \text{H=Sp(2n)} \end{cases}$$

$$\Phi_H^+ = \begin{cases} \{(x_i \pm x_j) | i < j\} & \text{H= 0(2n)} \\ \{(x_i \pm x_j), 2x_i | i < j\} & \text{H=Sp(2n)} \end{cases}$$

and in either case

$$\Phi_M = \{\pm(x_i - x_j) | i < j\}$$

$$\Phi_M^+ = \{(x_i - x_j) | i < j\} .$$

As the set of simple roots Δ_H we have

$$\Delta_H = \begin{cases} \{\alpha_i = x_i - x_{i+1} | i = 1, \cdots, n-1\} \cup \{\alpha_n = x_{n_1} + x_n\} & H = 0(2n) \\ \{\alpha_i = x_i - x_{i+1} | i = 1, \cdots, n-1\} \cup \{\alpha_n = 2x_n\} & \text{H=Sp(2n)} \end{cases} .$$

Let $W_H = N_H(T)/C_H(T)$ be the Weyl group of T in H. If $H = Sp(2n)$ then W_H is the same as the Weyl group $W(\Phi_H)$ of the root system Φ_H. However, if $H = 0(2n)$ then W_H is the semi-direct product of the Weyl group $W(\Phi_H)$ of the root system and the group of order two generated by the diagram automorphism w' which interchanges α_{n-1} and α_n, i.e., $W_H = W(\Phi_H) \quad < w' >$. In either case the order of W_H is $2^n n!$. $W_M = N_M(T)/C_M(T)$ is the Weyl group of T in M and it agrees with the Weyl group $W(\Phi_M)$ of the root system Φ_M. The order of W_M is $n!$. Let Ω be the distinguished set of coset representatives for $W_M \backslash W_H$ obtained by choosing the unique element of minimal length in each coset. (For the case $H = O(2n)$ the length function on $W(\Phi_H)$ is extended to W_H by setting $\ell(w') = 0$.) The following lemma gives a description of Ω.

Lemma 5.1: The elements $w \in \Omega$ are precisely those Weyl group elements given by

$$w^{-1}: \quad \begin{array}{ccc} x_1 & \mapsto & x_{i_1} \\ & \vdots & \\ x_k & \mapsto & x_{i_k} \\ x_{k+1} & \mapsto & -x_{i_{k+1}} \\ & \vdots & \\ x_n & \mapsto & -x_{i_n} \end{array}$$

with $i_1 < i_2 < \cdots < i_k$ and $i_{k+1} > \cdots > i_n$.

Proof: It is easy to see that these elements are indeed in W_H and that there are $2^n = |W_M \setminus W_H|$ of them. By Proposition 1.1.3 of Casselman [C] the unique element of minimal length in each coset is characterized by $w^{-1}\Phi_M^+ \subset \Phi_H^+$. (There is no problem in extending this result to the disconnected $O(2n)$ case.) It is an easy enough matter to check that the elements listed in the lemma have this property.

Each element $w \in \Omega$ gives a decomposition of N as follows. To each $\alpha \in \Phi_H^+$ there is associated a one parameter unipotent subgroup $N_\alpha \subset N$. For $w \in \Omega$ we then set

$$N_w^- = \Pi_{\substack{\alpha>0 \\ w\alpha<0}} N_\alpha$$

$$N^w = w^{-1}Pw \cap N .$$

Then for each w we have $N = N^w N_w^-$.

For each finite place v of k, let K_v be the maximal compact subgroup of H_v given by $K_v = H_v \cap GL(2n, \mathcal{O}_v)$. For v an infinite place, take K_v to be any maximal compact. Then at each place we have $H_v = P_v K_v = B_v K_v$ and $H = P_{\mathbb{A}} K_{\mathbb{A}} = B_{\mathbb{A}} K_{\mathbb{A}}$ with $K_{\mathbb{A}} = \Pi_v' K_v$.

To unify the formulas for $O(2n)$ and $Sp(2n)$ in what follows, we set

$$\nu_H = \begin{cases} 0 & \mathrm{H} = O(2n) \\ 1 & \mathrm{H} = Sp(2n) \end{cases} .$$

5.2. Let us now turn to the construction of the Eisenstein series. Let δ_p denote the modulus character of P. If $p = mu \in P_{\mathbb{A}}$ with $m \in M_{\mathbb{A}}$ and $n \in U_{\mathbb{A}}$, then

$\delta_P(P) = |det(m)|$. For $s \in \mathbb{C}$, set $\delta_P^s(p) = |det(m)|^s$. Let $\Phi'_{K,s}(h)$ be the extension of δ_P^s to H given by $\Phi_{K,s}(m\ u\ k) = \delta_P^s(m)$, $k \in K_{\mathbb{A}}$. Then $\Phi'_{K,s} \in ind_{P_{\mathbb{A}}}^{H_{\mathbb{A}}}(\delta_P^s)$ and $\Phi'_{K,s}(1) = 1$. Let ρ_s denote the action of $H_{\mathbb{A}}$ on $ind_P^H(\delta_P^s)$ by right translation. If $\mathcal{H}_{\mathbb{A}}$ is the adelic Hecke algebra of $H_{\mathbb{A}}$ then the representation ρ_s defines a representation of $\mathcal{H}_{\mathbb{A}}$ on $ind_P^H(\delta_P^s)$, which will also be denoted by ρ_s, by right convolution. So for $f \in ind_P^H(\delta_P^s)$ and $\beta \in \mathcal{H}_{\mathbb{A}}$, $\phi_s(\beta)f = f * \beta$. Let $I_s \subset ind_P^H(\delta_P^s)$ be the subspace spanned by the functions of the form $\Phi'_{K,s} * \beta$ with $\beta \in \mathcal{H}_{\mathbb{A}}$.

The family of Eisenstein series in which we are interested are those formed from the functions in I_s. So for $f \in I_s$ we set

(5.1) $$E_f(h;s) = \sum_{\gamma \in P_k \backslash H_k} f(\gamma h)\ .$$

This will converge for $Re(s)$ large enough. To analyse the analytic behaviour of the $E_f(h;s)$ we will apply Langland's theory of the constant term.

First notice that the Eisenstein series are actually associated to representations induced from $B_{\mathbb{A}}$. Let δ_B be the modulus character of B. It is the extension to B of the rational character of T given by the root $2\rho_B = \sum_{\alpha \in \Phi^+(H)} \alpha = \sum_{i=1}^n (n - i + \nu_H)x_i$. For χ any character of $T_k \setminus T_{\mathbb{A}}$, extended to a character of $B_{\mathbb{A}}$, define the (normalized) induced representation by $Ind_{B_{\mathbb{A}}}^{H_{\mathbb{A}}}(\chi) = \{f : H_{\mathbb{A}} \to \mathbb{C} \,|\, f(bh) = \chi(b)\delta_B(b)^{1/2} f(h)$ for $\mathrm{b} \in \mathrm{B}_{\mathbb{A}}$ and $\mathrm{h} \in \mathrm{H}_{\mathbb{A}}\}$. $H_{\mathbb{A}}$ acts on this space by right translation as usual.

<u>**Lemma 5.2:**</u> For $s \in \mathbb{C}$, let χ_s be the character of $T_k \setminus T_{\mathbb{A}}$ defined by $\chi_s = \sum_{i=1}^n (s + i - n - \nu_H)x_i \in X^*(T) \otimes_{\mathbb{Z}} \mathbb{C}$. Then $ind_{P_{\mathbb{A}}}^{H_{\mathbb{A}}}(\delta_P^s) \subset Ind_{B_{\mathbb{A}}}^{H_{\mathbb{A}}}(\chi_s)$.

<u>**Proof:**</u> For $b \in B_{\mathbb{A}}$, write $b = \underline{t}n$ with $\underline{t} \in T_{\mathbb{A}}$ and $n \in N_{\mathbb{A}}$. We may further decompose $n = mu$ with $m \in N_{\mathbb{A}} \cap M_{\mathbb{A}}$ and $u \in U_{\mathbb{A}}$. Then for $f \in ind_{P_{\mathbb{A}}}^{H_{\mathbb{A}}}(\delta_P^s)$ we have $f(bh) = f(\underline{t}m\ u\ h) = \delta_P^s(\underline{t}m)f(h) = |det(\underline{t}m)|^s f(h) = |t_1 \cdots t_n|^s f(h)$. Therefore, under $B_{\mathbb{A}}$, f transforms according to the character $\alpha_s = \sum_{i=1}^n s x_i \in X^*(T) \otimes_{\mathbb{Z}} \mathbb{C}$. If we then write $\alpha_s = \chi_s + \rho_B$ then $\chi_s = \sum_{i=1}^n (s + i - n - \nu_H)x_i$.

Since the Eisenstein series $E_f(h;s)$ are associated to a representation induced from $B_{\mathbb{A}}$, the constant term to study is that along $N_{\mathbb{A}}$. Let

(5.2) $$E_f^\circ(h;s) = \int_{N_k \backslash N_{\mathbb{A}}} E_f(nh;s)dn\ .$$

According to Langlands [L2], $E_f(h;s)$ and $E_f^\circ(h;s)$ enjoy the same analytic properties in s, that is, analytically continue to the same regions in the s-plane and satisfy the same functional equation.

To analyse $E_f^\circ(h;s)$, insert the definition of $E(h;s)$ in (5.1) into (5.2) and unfold to obtain

$$E_f^\circ(h;s) = \sum_{\gamma \in P_k \backslash H_k / N_k} \int_{N_k^\gamma \backslash N_{\mathbb{A}}} f(\gamma\ n\ h)dn$$

where $N^\gamma = \gamma^{-1} P \gamma \cap N$. Since δ_P^s is trivial on $P_{\mathbb{A}} \cap \gamma N_{\mathbb{A}} \gamma^{-1}$ we have

$$E_f^\circ(h;s) = \sum_{\gamma \in P_k \backslash H_k / N_k} \int_{N_{\mathbb{A}}^\gamma \backslash N_{\mathbb{A}}} f(\gamma\ n\ h)dn\ . \tag{5.3}$$

(The measure on $N_{\mathbb{A}}$ is taken so that all of the volumes $vol(N_k^\gamma \setminus N_{\mathbb{A}}^\gamma) = 1$.) By the relative Bruhat theory, each double coset $P_k \setminus H_k / N_k$ has a representative in W_H and two elements of W_H represent the same coset iff they are in the same left W_M coset, i.e., $P_k \setminus H_k / N_k = W_M \setminus W_H$. Recalling our choice Ω of coset representatives for $W_M \setminus W_H$, (5.3) becomes

$$\begin{aligned} E_f^\circ(h;s) &= \sum_{w \in \Omega} \int_{N_{\mathbb{A}}^w \backslash N_{\mathbb{A}}} f(w\ n\ h)dn \\ & \sum_{w \in \Omega} \int_{N_{w,\mathbb{A}}^-} f(w\ n\ h)dn\ . \end{aligned} \tag{5.4}$$

As usual the integral

$$I_w(f)(h) = \int_{N_{w,\mathbb{A}}^-} f(w\ n\ h)dn \tag{5.5}$$

defines an intertwining operator $I_w : Ind_{B_{\mathbb{A}}}^{H_{\mathbb{A}}}(\chi_s) \longrightarrow Ind_{B_{\mathbb{A}}}^{H_{\mathbb{A}}}(w^{-1} \cdot \chi_s)$, where W_H acts on the characters $\chi \in X^*(T) \otimes \mathbb{C}$ by $w \cdot \chi(\underline{t}) = \chi(w^{-1} \underline{t} w)$. In each $Ind_{B_{\mathbb{A}}}^{H_{\mathbb{A}}}(\chi)$ there is a unique K-fixed vector, $\Phi_{K,\chi}(h)$ normalized so that $\Phi_{K,\chi}(1) = 1$. We will set $\Phi_{K,s}(h) = \Phi_{K,\chi_s}(h)$ and $\Phi_{K,w^{-1}\cdot s}(h) = \Phi_{K,w^{-1}} \cdot \chi_s(h)$. If we set

$$c_w(s) = \int_{N_{w,\mathbb{A}}^-} \Phi_{K,s}(w\ n)dn = I_w(\Phi_{K,s})(1)$$

then we have $I_w(\Phi_{K,s}) = c_w(s) \Phi_{K,w^{-1}\cdot s}$.

Returning to our constant term, (5.4) can now be rewritten as

$$E_f^\circ(h;s) = \sum_{w \in \Omega} I_w(f)(h)\ . \tag{5.6}$$

Hence the analytic properties of $E_f^\circ(h;s)$ are reduced to the analytic properties of the $I_w(f)$. For $f \in \mathcal{I}_s$ we can further reduce to just considering the analytic properties of $c_w(s)$.

Proposition 5.1: For $f \in \mathcal{I}_s$ and $w \in \Omega$, $I_w(f)$ and $c_w(s)$ have the same set of poles.

Proof: Since $ind_{P_{\mathbb{A}}}^{H_{\mathbb{A}}}(\delta_P^s) \subset Ind_{B_{\mathbb{A}}}^{H_{\mathbb{A}}}(\chi_s)$, $\Phi'_{K,s} = \Phi_{K,s}$ and $\mathcal{I}_s$ is the space spanned by the functions $\Phi_{K,s} * \beta$ with $\beta \in \mathcal{H}_{\mathbb{A}}$. It will be enough to consider the $\Phi_{K,s} * \beta$.

First, assume $f = \Phi_{K,s}$, so that $I_w(f) = c_w(s)\Phi_{K,w^{-1}\cdot s'}$. Since $\Phi_{K,w^{-1}\cdot s}$ is K-invariant, it is completely determined by its restriction to $B_{\mathbb{A}}$. However, on $B_{\mathbb{A}}$ it restricts to the character χ_s which is an entire function of s. So $\Phi_{K,w^{-1}} \cdot s$ is entire as a function of s. Hence $I_w(\Phi_{K,s})$ and $c_w(s)$ have the same set of poles.

If $f = \Phi_{K,s} * \beta$, then since I_w intertwines the representations of $\mathcal{H}_{\mathbb{A}}$ we have $I_w(f) = I_w(\Phi_{K,s} * \beta) = I_w(\Phi_{K,s}) * \beta = c_w(s)\Phi_{K,w^{-1}\cdot s} * \beta$. Since $\Phi_{K,w^{-1}\cdot s}$ is entire, so will $\Phi_{K,w^{-1}\cdot s} * \beta$ be. Hence, again, $I_w(f)$ and $c_w(s)$ have the same set of poles.

From the proof of Proposition 5.1 we record the following corollary for future use.

Corollary: For $f = \Phi_{K,s} * \beta \in \mathcal{I}_s$,

$$E_f^\circ(h;s) = \sum_{w\in\Omega} c_w(s)(\Phi_{K,w^{-1}\cdot s} * \beta)(h) .$$

5.3. By the corollary to proposition 5.1, the possible poles of $E_f(h;s)$ are determined by the poles of the $c_w(s)$. To compute the $c_w(s)$ we first use the method of Gindikin and Karpelevich. For each root $\alpha \in \Phi_H$, there is a co-root $\overset{v}{\alpha} \in X^*(T) \otimes_{\mathbb{Z}} \mathbb{Q}$ defined by $\overset{v}{\alpha} = \alpha/2(\alpha,\alpha)$, and the $\overset{v}{\alpha}$ make up the co-root system Φ_H^v . For each half space $R \subset X^*(T) \otimes_{\mathbb{Z}} \mathbb{Q}$ bounded by a hyperplane passing through zero let $\Phi_H^-(R)$ be the set of negative roots lying in R and $N(R) \subset N^-$ the unipotent subgroup generated by the root subgroups corresponding to the roots in $\Phi_H^-(R)$. So $N(R) = \Pi_{\alpha\in\Phi_H^-(R)} N_\alpha$. Then the method of Gindikin-Karpelevich gives the following.

Proposition 5.2: For $\mu \in X^*(T) \otimes_{\mathbb{Z}} \mathbb{C}$,

$$\int_{N(R)_{\mathbb{A}}} \Phi_{K,\mu}(n)\,dn = \prod_{-\alpha\in\Phi_H^-(R)} \frac{\zeta((\mu,\alpha^v))}{\zeta((\mu,\alpha^v)+1)}$$

where ς is the zeta function of the field k together with its gamma factors.

Since the integral factors as a product of local integrals, it is enough to prove the corresponding local statement. This was done for $k = \mathbb{R}$ by Gindikin and Karpelevich [G-K] and their method was used by Langlands[L1] to prove the formula for $k = \mathbb{Q}_p$. The result for arbitrary local field k goes the same way.

By lemma 5.1, for each $w \in \Omega$ there is an integer, which we will denote by k_w, such that $w^{-1}x_i \in \{x_1, \cdots, x_n\}$ for $i \leq k_w$ and $w^{-1}x_i \in \{-x_1, \cdots, -x_n\}$ for $i > k_w$. Each $w \in \Omega$ then defines a bijective map $i_w : \{1, \cdots, n\} \longrightarrow \{1, \cdots, n\}$ by $w^{-1}x_r = x_{i_w(r)}$ if $1 \leq r \leq k_w$ and $w^{-1}x_r = -x_{i_w(r)}$ if $k_w < r \leq n$. As a notational convenience, let us put $[a, b]_{\mathbb{Z}} = \{n \in \mathbb{Z} : a \leq n \leq b\}$. Then $i_w : [1, n]_{\mathbb{Z}} \longrightarrow [1, n]_{\mathbb{Z}}$ is increasing on $[1, k_w]_{\mathbb{Z}}$ and decreasing on $[k_w + 1, n]_{\mathbb{Z}}$. Applying the method of Gindikin-Karpelevich then gives the following.

Proposition 5.3: For $w \in \Omega$,

$$c_w(s) = c_w^0(s) c_w^i(s) c_w^r(s) ,$$

where

$$c_w^0(s) = \begin{cases} 1 & \text{H=0(2n)} \\ \frac{\varsigma(s-n+k_w)}{\varsigma(s)} & \text{H=Sp(2n)} \end{cases}$$

$$c_w^i(s) = \prod_{\substack{\ell \leq k_w < m \\ i_w(m) > i_w(\ell)}} \frac{\varsigma(2s - 2n - 2\nu_H + \ell + m)}{\varsigma(2s - 2n - 2\nu_H + \ell + m + 1)}$$

$$c_w^r(s) = \prod_{k_w < \ell < m \leq n} \frac{\varsigma(2s - 2n - 2\nu_H + \ell + m)}{\varsigma(2s - 2n - 2\nu_H + \ell + m + 1)} .$$

Pf: Recall that $c_w(s) = \int_{N^-_{w,\mathbb{A}}} \Phi_{K,s}(wn)dn$. Since $w \in K$ and $\Phi_{K,s}$ is K-invariant, we have

$$c_w(s) = \int_{N^-_{w,\mathbb{A}}} \Phi_{K,s}(w\ n\ w^{-1})dn = \int_{wN^-_{w,\mathbb{A}}} w^{-1}\Phi_{K,s}(n)dn .$$

From the definition of N_w^- , it is easy to see that $wN_w^-w^{-1} = \Pi_{\substack{w^{-1}\alpha>0 \\ \alpha<0}} N_\alpha$. Now let E be any half space containing Φ_H^+ and let $R = wE$. Then $\Phi_H^-(R) = \{\alpha \in \Phi_H : \alpha < 0$ and $w^{-1}\alpha > 0\}$ and hence $wN^-_{w,\mathbb{A}}w^{-1} = N(R)_{\mathbb{A}}$. Applying proposition 5.2 then gives

$$c_w(s) = \prod_{\alpha \in -\Phi_H^-(R)} \frac{\varsigma((\chi_s, \check{\alpha}))}{\varsigma((\chi_s, \check{\alpha}) + 1)} .$$

To complete the proposition, we must determine the roots $\alpha \in -\Phi_H^-(R) = \{\alpha \in \Phi_H | \alpha > 0 \text{ and } w^{-1}\alpha < 0\}$. If we let

$$A_w^r = \{x_\ell + x_m | k_w < \ell < m \le n\}$$

$$A_w^i = \{x_\ell + x_m | \ell \le k_w < m \quad \text{and} \quad \mathrm{i_w}(\ell) < \mathrm{i_w}(\mathrm{m})\}$$

and

$$A_w^0 = \begin{cases} \emptyset & \mathrm{H{=}0(2n)} \\ \{2x_i | k_w < i \le n\} & \mathrm{H = Sp(2n)} \end{cases}$$

then using lemma 5.1a it is easy to compute that $\{\alpha \in \Phi_H | \alpha > 0 \text{ and } w^{-1}\alpha < 0\} = A_w^0 \cup A_w^i \cup A_w^r$. Since the x_i satisfy $(x_i, x_j) = \delta_{ij}$ we have

$$(x_\ell + x_m)^v = x_\ell + x_m$$

$$(2x_i)^v = x_i$$

$$(\chi_s, x_\ell + x_m) = 2s - 2n - 2\nu_H + \ell + m$$

$$(\chi_s, x_i) = s + i - n - \nu_H \ .$$

Then we have

$$\prod_{\alpha \in A_w^r} \frac{\varsigma((\chi_s, \overset{v}{\alpha}))}{\varsigma((\chi_s, \overset{v}{\alpha}) + 1)} = c_w^r(s)$$

$$\prod_{\alpha \in A_w^i} \frac{\varsigma((\chi_s, \overset{v}{\alpha}))}{\varsigma((\chi_s, \overset{v}{\alpha}) + 1)} = c_w^i(s)$$

and

$$\prod_{\alpha \in A_w^0} \frac{\varsigma((\chi_s, \overset{v}{\alpha}))}{\varsigma((\chi_s, \overset{v}{\alpha} + 1)} = c_w^0(s)$$

where we always take empty products to be $\equiv 1$.

We will henceforth call $c_w^r(s)$ the regular term in $c_w(s)$ and $c_w^i(s)$ the irregular term. The regular term easily simplifies. For a fixed value of ℓ, the m variable will run from $\ell + 1$ to n. Thus

(5.7)
$$\begin{aligned} c_w^r(s) &= \prod_{\ell=k_w+1}^{n-1} \left(\prod_{m=\ell+1}^{n} \frac{\varsigma(2s - -2n - 2\nu_H + \ell + m)}{\varsigma(2s - 2n - 2\nu_H + \ell + m + 1)} \right) \\ &= \prod_{\ell=k_w+1}^{n-1} \frac{\varsigma(2s - 2n - 2\nu_H + 2\ell + 1)}{\varsigma(2s - n - 2\nu_H + \ell + 1)} \ . \end{aligned}$$

Hence the regular term depends only upon k_w .

To simplify the irregular term, we must know more about the combinatorics of the action of w^{-1} on the roots, i.e., of the map i_w .

Lemma 5.3: For $r \in [1,n]_{\mathbb{Z}}$, let $M_w(r) = \{m | k_w < m \leq n \text{ and } i_w(m) < i_w(r)\}$. Then $i_w(n)$ is the minimal $r \in [1,k_w]_{\mathbb{Z}}$ with $M_w(r) \neq \emptyset$.

Proof: If $i_w(n) = 1$, then $i_w(1) \in [2,n]_{\mathbb{Z}}$ and hence $i_w(1) > 1$. Therefore $n \in M_w(1)$ and $M_w(1) \neq \emptyset$.

Next suppose $i_w(n) = a > 1$. Since i_w is increasing on $[1,k_w]_{\mathbb{Z}}$ and decreasing on $[k_w+1,n]_{\mathbb{Z}}$ we see that we must have $i_w(t) = t$ for $t \leq a-1$, and hence $M_w(t) = \emptyset$ for $t \leq a-1$. On the other hand, since $i_w(n) = a$ we must have $i_w(a) > a$. Therefore $n \in M_w(a)$. This proves the lemma.

Since i_w is increasing on $[1,k_w]_{\mathbb{Z}}$, we will have $M_w(r) \neq \emptyset$ for all $r \in [i_w(n), n-1]_{\mathbb{Z}}$. For $r \in [i_w(n), n-1]_{\mathbb{Z}}$ we let $\mu_w(r)$ be the minimal element of $M_w(r)$. Then in the formula for the irregular term in proposition 5.3, the ℓ parameter will run over the values $i_w(n) \leq \ell \leq k_w$ and, for a fixed value of ℓ, m will run over $\mu_w(\ell) \leq m \leq n$, since i_w is decreasing on $[k_w+1,n]_{\mathbb{Z}}$. So we get

$$(5.8)\qquad \begin{aligned} c^i_w(s) &= \prod_{\ell=i_w(n)}^{k_w} \left(\prod_{m=\mu_w(\ell)}^{n} \frac{\varsigma(2s-2n-2\nu_H+\ell+m)}{\varsigma(2s-2n-2\nu_H+\ell+m+1)} \right) \\ &= \prod_{\ell=i_w(n)}^{k_w} \frac{\varsigma(2s-2n-2\nu_H+\ell+\mu_w(\ell))}{\varsigma(2s-n-2\nu_H+\ell+1)} . \end{aligned}$$

We record these simplifications in the following corollary to proposition 5.3.

Corollary 1: For $w \in \Omega$, $c_w(s) = c^0_w(s)c^i_w(s)c^r_w(s)$ with $c^0_w(s)$ as in proposition 5.3 and

$$c^i_w(s) = \prod_{\ell=i_w(n)}^{k_w} \frac{\varsigma(2s-2n-2\nu_H+\ell+\mu_w(\ell))}{\varsigma(2s-n-2\mu_H+\ell+1)}$$

$$c^r_w(s) = \prod_{\ell+k_w+1}^{n-1} \frac{\varsigma(2s-2n-2\nu_H+2\ell+1)}{\varsigma(2s-n-2\nu_H+\ell+1)}$$

In what follows, it will be useful to have another expression for $c_w(s)$ as well. This will be derived from the expression in corollary 1 through the following series of lemmas.

Lemma 5.4: For $0 \leq j \leq n - k_w - 1$ let

$$I_w(j) = \begin{cases} [i_w(n-j), i_w(n-j-1) - j - 2]_{\mathbb{Z}} & 0 \leq j \leq n - k_w - 2 \\ [i_w(n-j) - j, k_w]_{\mathbb{Z}} & j = n - k_w - 1 \end{cases} .$$

Then for $\ell \in [i_w(n), k_w]_{\mathbb{Z}}$ we have $\ell \in I_w(j)$ iff $\mu_w(\ell) = n - j$.

Proof: The possible values for $\mu_w(\ell)$ are $\{k_w + 1, \cdots, n\}$. Since $[k_w + 1, n]_{\mathbb{Z}}$ decomposes disjointly as $[k_w + 1, n]_{\mathbb{Z}} = \amalg_{j=0}^{n-k_w-2} I_w(j)$ it is enough to show that $\mu_w(\ell) = n - j$ implies $\ell \in I_w(j)$.

Since i_w is increasing on $[1, k_w]_{\mathbb{Z}}$ and decreasing on $[k_w+1, n]_{\mathbb{Z}}$ it is easy to see that for $\ell \in [1, k_w]_{\mathbb{Z}}$ we have $i_w(\ell) = \ell + |M_w(\ell)|$. If $\ell \in [1, i_w(n) - 1]_{\mathbb{Z}}$, then $i_w(\ell) = \ell$. However, if $\ell \geq i_w(n)$, so that $M_w(\ell) \neq \emptyset$, then $i_w(\ell) = \ell + n - \mu_w(\ell) + 1$. Now, for $\ell \in [i_w(n), k_w]$, we have by definition that $\mu_w(\ell) = n - j$ implies $i_w(n-j) < i_w(\ell) < i_w(n-j-1)$ for $0 \leq j \leq n - k_w - 2$. Using the above formula for $i_w(\ell)$, this gives $\ell \in I_w(j)$ if $\mu_w(\ell) = n - j$. On the other hand, if $\mu_w(\ell) = k_w + 1$, then by definition $i_w(k_w + 1) < i_w(\ell)$ and we get $\ell \in I_w(n - k_w - 1)$. This proves the lemma.

Lemma 5.5: The irregular term $c^i_w(s)$ in $c_w(s)$ can be rewritten as $c^i_w(s) = a^i_w(s) b^i_w(s)^{-1}$ where

$$a^i_w(s) = \prod_{j=0}^{n-k_w-1} \varsigma(2s - n - 2\nu_H + i_w(n-j) - 2j)$$

and

$$b^i_w(s) = \prod_{\ell=-n+2k_w+2}^{k_w+1} \varsigma(2s - n - 2\nu_H + \ell) .$$

Proof: We begin with the formula (5.8) for the irregular term. The interval $[i_w(n), k_w]_{\mathbb{Z}}$ has a disjoint decomposition $[i_w(n), k_w]_{\mathbb{Z}} = \amalg_{j=0}^{n-k_w-1} I_w(j)$, where the $I_w(j)$ are as in lemma 5.4. Applying lemma 5.4 to (5.8) we then get

$$c^i_w(s) = \prod_{j=0}^{n-k_w-1} \left(\prod_{\ell \in I_w(j)} \frac{\varsigma(2s - n - 2\nu_H + \ell - j)}{\varsigma(2s - n - 2\nu_H + \ell + 1)} \right) .$$

As a simple exercise in cancellation, we have the formula

$$\prod_{\ell=a}^{b} \frac{(A + \ell - j)}{(A + \ell + 1)} = \left(\prod_{\ell=a-j}^{a} \varsigma(A + \ell) \right) \left(\prod_{\ell=b+1-j}^{b+1} \varsigma(A + \ell) \right)^{-1}$$

when $a \le b+1$, with empty products taken to be equal to 1 . If we apply this with $A = 2s - n - 2\nu_H$ to each of the inner products and let

$$I_{w,a}(j) = [i_w(n-j) - 2j, i_w(n-j) - j]_{\mathbb{Z}}$$

and

$$I_{w,b}(j) = \begin{cases} [i_w(n-j-1) - 2j - 1, i_w(n-j-1) - j - 1]_{\mathbb{Z}} & 0 \le j \le n - k_w - 2 \\ [-n + 2k_2 + 2, k_w + 1]_{\mathbb{Z}} & j = n - k_w - 1 \end{cases}$$

then we have

$$c_w^i(s) = \prod_{j=0}^{n-k_w-1} \left(\prod_{\ell \in I_{w,a}(j)} \varsigma(2s - n - 2\nu_H + \ell) \right) \left(\prod_{\ell \in I_{w,b}(j)} \varsigma(2s - n - 2\nu_H + \ell) \right)^{-1}$$

If we now separate the numerator and denominator, reindex the denominator and cancel common terms, we get the lemma.

If we now take the expression for the irregular term $c_w^i(s)$ from lemma 5.5 and the expression for the regular term $c_w^r(s)$ from (5.7), multiply then together, and cancel common terms from the numerator and denominator of the resulting expression, we arrive at the following corollary of proposition 5.3.

<u>**Corollary 2:**</u> For $w \in \Omega$, $c_w(s) = c_w^0(s) a_w(s) b_w(s)^{-1}$ with $c_w^0(s)$ as in proposition 5.3 and

$$a_w(s) = \prod_{j=0}^{n-k_w-1} \varsigma(2s - n - 2\nu_H + i_w(n-j) - 2j)$$

and

$$b_w(s) = \prod_{j+\nu_H}^{n+\nu_H-k_w-1} \varsigma(2s - 2j) \ .$$

5.4. There is a finite product of zeta functions which will clear the denominators of all the $c_w(s)$ and at the same time yield a nice functional equation for the constant term of the Eisenstein series. If is obtained from the denominator of $c_{w_0}(s)$, where w_0 is the longest element in Ω . It is easy to check from the description of the elements of Ω given in lemma 5.1 that w_0 is given by

(5.9)
$$w_0^{-1} : \begin{matrix} x_1 & \longrightarrow & -x_n \\ \vdots & & \vdots \\ x_n & \longrightarrow & -x_1 \end{matrix} \ .$$

(For $H = 0(2n)$ there are actually two longest elements, and either one will do, since they both give the same $c_{w_0}(s)$.)

Lemma 5.6: For $w_0 \in \Omega$ the longest element as in (5.9) we may write

$$c_{w_0}(s) = \frac{a^0_{w_0}(s)a^r_{w_0}(s)}{b^0_{w_0}(s)b^r_{w_0}(s)}$$

where

$$a^0_{w_0}(s) = \begin{cases} 1 & \text{H=0(2n)} \\ \varsigma(s_n) & \text{H=Sp(2n)} \end{cases}$$

$$b^0_{w_0}(s) = \begin{cases} 1 & \text{H=0(2n)} \\ \varsigma(s) & \text{H=Sp(2n)} \end{cases}$$

$$a^r(w_0(s) = \prod_{\substack{j=n+2\nu_H-1 \\ j\equiv 1(2)}}^{2n+2\nu_H-3} \varsigma(2s-j)$$

$$b^r_{w_0}(s) = \prod_{\substack{j=2\nu_H \\ j\equiv 0(2)}}^{n+2\nu_H-2} \quad .$$

Proof: For w_0 the longest root we have $k_{w_0} = 0$ and $i_{w_0}(n) = 1$. Therefore we have

$$c^0_{w_0}(s) = \begin{cases} 1 & \text{H=0(2n)} \\ \varsigma(s-n)\varsigma(s)^{-1} & \text{H=Sp(2n)} \end{cases}$$

$$c^r_{w_0}(s) = \prod_{\ell=1}^{n-1} \frac{\varsigma(2s-2n-2\nu_H+2\ell+1)}{\varsigma(2s-n-2\nu_H+\ell+1)}$$

$$c^i_{w_0}(s) = 1$$

from the corollary 1 to proposition 5.3. If we cancel the common zeta functions from the numerator and denominator of $c^r_{w_0}(s)$ we get

$$c^r_{w_0}(s) = a^r_{w_0}(s)b^r_{w_0}(s)^{-1}$$

with $a^r_{w_0}(s)$ and $b^r_{w_0}(s)$ as in the statement of the lemma.

We now set $d_H(s)$ to be the denominator $b^0_{w_0}(s)b^r_{w_0}(s)$ of $c_{w_0}(s)$. So

$$d_H(s) = \begin{cases} \prod^{n-2}_{\substack{j=0 \\ j\equiv 0(2)}} \varsigma(2s-j) & \text{H=0(2n)} \\ \varsigma(s)\prod^{n}_{\substack{j=2 \\ j\equiv 0(2)}} \varsigma(2s-j) & \text{H=Sp(2n)} \end{cases} .$$

As to the claim that $d_H(s)$ cancels the denominator of all $c_w(s)$ for $w \in \Omega$, we have the following lemma.

Lemma 5.7: For $w \in \Omega$ set

$$a_w^0(s) = \begin{cases} 1 & \text{H=0(2n)} \\ \varsigma(s-n+k_w) & \text{H=Sp(2n)} \end{cases} .$$

(a) If $n \geq 2k_w + 1$ then

$$d_H(s)c_w(s) = a_w^0(s) \prod_{j=n+2\nu_H-i_w(n)}^{n+2\nu_H-2} \varsigma(2s-j)$$

$$\prod_{\substack{j=n+2\nu_H-1 \\ j\equiv 1(2)}}^{2n+2\nu_H-2k_w-3} \varsigma(2s-j) \cdot \prod_{j=i_w(n)}^{k_w} \varsigma(2s-2n+j+\mu_w(j))$$

(b) If $n \leq 2k_w$ then

$$d_H(s)c_w(s) = a_w^0(s) \prod_{\substack{j=2n+2\nu_H-2k_w \\ j\equiv 0(2)}}^{n+2\nu_H-2} \varsigma(2s-j)$$

$$\prod_{j=0}^{n-k_w-1} \varsigma(2s-n-2\nu_H+i_w(n-j)-2j) .$$

Proof: To derive (a), take the formula for $c_w(s)$ given in corollary 1 to proposition 5.3, multiply by $d_H(s)$, and cancel common zeta functions from the numerator and denominator. For (b), we do the same except that now we use the formula for $c_w(s)$ from corollary 2 to proposition 5.3.

From the formulas in lemma 5.7 we may now conclude the following.

Proposition 5.4: Let

$$\gamma_H = \begin{cases} n-1 & \text{H=0(2n)} \\ n+1 & \text{H=Sp(2n)} \end{cases} .$$

Then for $w \in \Omega$, the only possible poles for $d_H(s)c_w(s)$ are the integral and half integral points of the interval $[0, \gamma_H]$. The order of each pole is finite.

Remark: This proposition can also be proven using the single formula for $c_w(s)$ given in corollary 1 to proposition 5.3 and the following combinatorial lemma. However, this will not yield explicit formulas for $d_H(s)c_w(s)$. We present the necessary lemma for $H = 0(2n)$, that for $Sp(2n)$ being similar.

<u>**A combinatorial lemma:**</u> Let $i : [1,n]_{\mathbb{Z}} \longrightarrow [1,n]_{\mathbb{Z}}$ be a bijection such that there is a k, $1 \le n \le n$, with i increasing on $[1,k]_{\mathbb{Z}}$ and decreasing on $[k+1,n]_{\mathbb{Z}}$ satisfying $n + i(n) + 1 \le 2k+1$. For $\ell \in [i(n),k]_{\mathbb{Z}}$ let $\mu(\ell) = min\{m | k < m \le n \text{ and } i(\ell) < i(m)\}$. Then every odd number occurring in $[n+i(n)+1, 2k+1]_{\mathbb{Z}}$ occurs also as a $\ell + \mu(\ell)$ for some $\ell \in [i(n),k]_{\mathbb{Z}}$.

Proof: The maximal value of k is $n-1$ and the minimal value of $i(n)$ is 1. So if $n + i(n) + 1 \le 2k+1$ we must have $3 \le n$.

We will proceed by induction on n . For $n = 3$ there is exactly one such map satisfying $n + i(n) + 1 \le 2k+1$ and that is $i(1) = 2$, $i(2) = 3$, $i(3) = 1$. So $k = 2$. Then $[n + i(n) + 1, 2k+1]_{\mathbb{Z}} = [5,5]_{\mathbb{Z}} \subset \{\ell + \mu(\ell) : 1 \le \ell \le 2\} = \{4,5\}$. So the lemma is true in this case.

Now assume $n \ge 4$. We proceed by cases.

(i) Assume $i(n) > 1$. Set $a = i(n) - 1$ and let $n' = n - a$. Define a bijection $i' : [1,n']_{\mathbb{Z}} \longrightarrow [1,n']_{\mathbb{Z}}$ by $i'(r) = i(r+a) - a$. Then the map i' satisfies the hypotheses of the lemma and has $n' < n$. So the lemma holds for i'. Since $k' = k - a$, we have $[n'+i'(n')+1, 2k'+1]_{\mathbb{Z}} = [n+i(n)+1, 2k+1]_{\mathbb{Z}} - 2a$. On the other hand $\mu'(\ell) = \mu(\ell+a) - a$ for $\ell \in [i'(n'),k']_{\mathbb{Z}}$. Therefore $\{\ell + \mu'(\ell) | i'(n') \le \ell \le k'\} = \{\ell + \mu(\ell) | i(n) \le \ell \le k\} - 2a$. So the conclusion for i is equivalent to the conclusion for i' , which is true by induction.

(ii) Assume $i(n) = 1$ and $i(n-1) = 1+a$ with $a \ge 2$. Then $\mu(\ell) = n$ for $1 \le \ell \le a-1$. So $\{\ell + \mu(\ell) | i(n) \le \ell \le k\} \supset \{\ell + n | 1 \le \ell \le a-1\} = [n+1, n+a-1]_{\mathbb{Z}}$. So it suffices to prove that the odd numbers in $[n+a, 2k+1]_{\mathbb{Z}}$ are contained in $\{\ell + \mu(\ell) | a \le \ell \le k\}$. Now set $n' = n - a$ and define a bijection $i' : [1,n']_{\mathbb{Z}} \longrightarrow [1,n']_{\mathbb{Z}}$ again by $i'(r) = i(r+a-1) - a$. This map satisfies the hypotheses of the lemma and hence the conclusion holds for i' by induction, since $n' < n$. For i' we have $k' = k - a + 1$, $\mu'(\ell) = \mu(\ell + a - 1) - a + 1$ and $i'(n') = 1$. Therefore $[n+a, 2k+1]_{\mathbb{Z}} = [n'+2, 2k'+1]_{\mathbb{Z}} + (2a-2)$ and $\{\ell + \mu(\ell) | a \le \ell \le k\} = \{\ell + \mu'(\ell) | 1 \le \ell \le kpr\} + (2a-2)$. Hence the conclusion for i follows from the conclusion for i', which is true by induction.

(iii) The final case is $i(n) = 1$ and $i(n-1) = 2$. So assume $i(n) = 1$, $i(n-1) = 2, \cdots, i(n - \alpha + 1) = \alpha$ with $\alpha \ge 2$ and $i(n-\alpha) - i(n\alpha + 1) = a \ge 2$. Then set $n' = n - \alpha + 1$ and define $i' : [1,n']_{\mathbb{Z}} \longrightarrow [1,n']_{\mathbb{Z}}$ by $i'(r) = i(r) - \alpha + 1$. Then i'

satisfies the hypotheses of the lemma and the lemma holds for i' by induction. We have $k' = k$, $i'(n') = 1$, and $\mu'(\ell) = \mu(\ell)$. So $\{\ell + \mu(\ell) | \leq \ell \leq k\} = \{\ell + \mu'(\ell) | 1 \leq \ell \leq k'\}$. However $[n' + i'(n') + 1,\ 2k' + 1]_{\mathbb{Z}} = [n + i(n) + 1 - \alpha,\ 2k + 1]_{\mathbb{Z}}$ and $[n + i(n) + 1 - \alpha,\ 2k + 1]_{\mathbb{Z}} \supset [n + i(n) + 1,\ 2k + 1]_{\mathbb{Z}}$. So, again, the lemma for i follows from that for i', which is true by induction.

This concludes the lemma.

This lemma just assures us that $d_H(s)$ will cancel the denominator of all the $c_w(s)$.

5.5. From the computation of the $c_w(s)$ given in section 5.3, we see that the Eisenstein series constructed in Section 5.2 may have an infinite number of poles. To obtain Eisenstein series with only a finite number of poles, we must use the factor $d_H(s)$ from Section 5.4 to modify our Eisenstein series.

Definition: For $f = \Phi_{K,s} * \beta \in I_s$, let

$$E_H(h; s; \beta) = d_H(s) E_f(h; s) \ .$$

With this modified Eisenstein series we now have the following theorem.

Theorem 5.1:

(a) $E_H(h; s; \beta)$ has a meromorphic continuation to the whole s-plane. Its poles can occur only at the integral and half integral points of $[0, \gamma_H]$.

(b) $E_H(h; s; \beta)$ satisfies the functional equation

$$E_H(h; s; \beta) = E_H(h; \gamma_H - s; \beta) \ .$$

Proof: From the work of Langlands (see[L2] or [A], for example) the analytic properties of $E_H(h; s; \beta)$ are determined by its constant term. Since $E_H^\circ(h; s; \beta) = d_H(s) E_f^\circ(h, s)$, then by the corollary to proposition 5.1 we have

$$E_H^\circ(h; s; \beta) = \sum_{w \in \Omega} d_H(s) c_w(s) \Phi_{K, w^{-1}s} * \beta) \ .$$

Since the $\Phi_{K,w^{-1}s} * \beta$ are entire functions of s ,as noted in the proof of proposition 5.1, the analytic properties of $E \circ_H (h; s; \beta)$ follow from those of the $d_H(s) c_w(s)$. These are given in proposition 5.4. This proves (a).

To prove (b), we again resort to the work of Langlands. Our notation is a little inconvenient for this, so let us put $E(h;f) = E_f(h;s)$. Then $E(h;f)$ satisfies the functional equations $E(h;f) = E(h;I_w(f))$ for $w \in \Omega$. (See, for example, [A]). We shall utilize this functional equation for $w = w_0$, the longest element in Ω. For $f = \Phi_{K,s} * \beta$, we have $I_{w_0}(f) = I_{w_0}(\Phi_{K,s}) * \beta = c_{w_0}(s)\Phi_{K,w_0^{-1}s} * \beta$. But a simple calculation shows that $w_0^{-1}\chi_s = \chi_{\gamma_H - s}$. Therefore $I_{w_0}(\Phi_{K,s} * \beta) = c_{w_0}(s)(\Phi_{K,\gamma_H - s} * \beta)$. Then we have

$$\begin{aligned} E_H(h;s;\beta) &= d_H(s)E\left(h;\Phi_{K,s} * \beta\right) \\ &= d_H(s)c_{w_0}(s)E\left(h;\Phi_{K,\gamma_H - s} * \beta\right) \\ &= \frac{d_H(s)}{d_H(\gamma_H - s)}c_{w_0}(s)E_H(h;\gamma_H - s;\beta) \end{aligned} .$$

If we apply the functional equation $\varsigma(s) = \varsigma(1-s)$ to $d_H(\gamma_H - s)$ we find that $c_{w_0}(s) = \frac{d_H(\gamma_H - s)}{d_H(s)}$. Therefore $E_H(h;s;\beta) = E_H(h;\gamma_H - s;\beta)$ and we are done.

§6. Computation of certain local factors for $O(2n)$ and $Sp(2n)$. Relation to Langlands L-functions.

In this section we would like to compute some of the local factors of the L-functions of Section 1. We will consider the case where G is either the split $O(2n)$ or $SP(2n)$. We will be interested in the local integrals

$$L_v(s;\varphi_{1,v},\varphi_{2,v},f_v) = d_{H_v}(s)\int_{G_v} f_v((g,1);s) < \pi(g)\varphi_{1,v},\varphi_{2,v} > dg$$

where $\varphi_{1,v}$ (resp. $\varphi_{2,v}$) is the K-fixed vector in an unramified principal series representation π_v (resp. $\tilde{\pi}_v$) and $f_v(h;s) = \Phi_{K_v,s}(h)$ is the normalized K-fixed vector of $ind_{P_v}^{H_v}(\delta_p^s)$. These are the local factors at the unramified places of the global L-function

$$L(s;\varphi_1,\varphi_2,f) = \int_{(G\times G)_k\backslash(G\times G)_{\mathbb{A}}} E_H((g_1,g_2)s;)\varphi_1(g_1)\varphi_2(g_2)dg_1dg_2$$

formed with the special Eisenstein series $E_H(h;s)$ constructed in Section 5 for $f(h;s) = \Phi_{K,s}(h)$.

As an application of these computations we will show that the Langlands L-functions$L(s,\pi,r)$ of π associated to the standard representation of r of the L-group ${}^LG$ of G has only a finite number of poles.

In the last section we analyze the special function $\Phi_{K,s}(g,1)$ as a function on G and derive an explicit formula for this function.

<u>6.1.</u> The computation of the local factors for $0(2n)$ and $Sp(2n)$ will proceed via a reduction to the local factors of $GL(n)$ sitting inside $0(2n)$ or $Sp(2n)$ as the Levi factor of a certain parabolic. In this first section we will compute the local factors for $GL(n)$ since they are of interest in their own right.

Fix k a non-archimedean local field and set $G = GL(n,k)$. Let $H = GL(2n,k)$ and embed $G\times G \hookrightarrow H$ as in Section 4. Then G acts on the vector space V and H on $W = V\oplus V$. Let $P\subset H$ be the parabolic subgroup stabilizing V^d, so that $P\cap G\times G = G^d$. Let τ be an irreducible unramified spherical representation of G and $\tilde{\tau}$ its contragredient. Let $v_0\in\tau$ be a K fixed vector and $\tilde{v}_0\in\tilde{\tau}$, also K-fixed, so that $< v_0,\tilde{v}_0 >= 1$. Here

K is the usual maximal compact subgroup of $GL(n)$. Let $f_0(h;s)$ be a K fixed vector in $ind_P^H(\delta_P^s)$. Then the local factors we are interested in computing are

$$L(s;v_0,\tilde{v}_0,f_0) = \int_G f_0((g,1);s) < \tau(g)v_0,\tilde{v}_0 > dg$$

for a particular choice of f_0 .

Write $W = V^+ \oplus V^-$ to distinguish the copies of V in W . We will write elements of H in block form according to this decomposition. Let $P' \subset H$ be the parabolic subgroup of H stabilizing V^- . If we set

$$A = \begin{pmatrix} 0 & 1 \\ 1 & 0 \end{pmatrix} \begin{pmatrix} 1 & 1 \\ 0 & 1 \end{pmatrix}$$

then A conjugates P into P' , i.e., $P' = APA^{-1}$. Consider $M(n,2n;k)$ as a right H-module and write elements $Z \in M(n,2n;k)$ as $Z = (Z_1,Z_2)$ with $Z_i \in M(n;k)$. If $\Phi \in S(M(n,2n;k))$ then we can write the elements of $ind_{P'}^H(\delta_{P'}^s)$ in the integral form

$$F'(h;\Phi) = |det\ h|^s \int_{GL(n)} \Phi((0,Z)h|det\ Z|^{2s}d^*Z\ .$$

If

$$p' = \begin{pmatrix} a & * \\ 0 & b \end{pmatrix} \quad \in P'$$

then $F'(p'h;\Phi) = |\frac{det\ (a)}{det\ (b)}|^s F'(h;\Phi)$. Since A conjugates P into P' , we obtain the elements of $ind_P^H(\delta_P^s)$ as

$$F(h;\Phi) = F'(Ah;\Phi) = |det\ h|^s \int_{GL(n)} \Phi((Z,Z)h)|det\ Z|^{2s}d^*Z\ .$$

If we take Φ_0 to be the characteristic function of $M(n,2n;\sigma)$, where σ is the ring of integers of k, then $F(h;\Phi_0)$is a K-fixed vector in $ind_P^H(\delta_P^s)$.

<u>**Proposition 6.1:**</u> Let τ, v_0 and $\tilde{v}_0$ be as above and let $f_0(h;s) = F(h;\Phi_0)$. Set $\omega(g) =< \tau(g)v_0,\tilde{v}_0 >$ and $\tilde{\omega}(g) =< v_0,\tilde{\tau}(g)\tilde{v}_0 >$. Then

$$L(s,v_0,\tilde{v}_0,f_0) = Z(\Phi_0,s,\omega)Z(\Phi_0,s,\tilde{\omega})$$

where $Z(\Phi_0,s,\omega)$ and $Z(\Phi_0,s,\tilde{\omega})$ are the local zeta integrals of Godement and Jacquet.

Proof. For convenience, let us temporarily set $L(s) = L(s, v_0, \tilde{v}_0, f_0)$. Then by definition

$$L(s) = \int_G f_0((g,1);s) < \tau(g)v_0, \tilde{v}_0 > dg$$
$$= \int_G |det\ g|^s \int_{GL(n)} \Phi_0((Zg, Z))|det\ Z|^{2s} d^*Z\omega(g)dg$$
$$= \int_G \int_{GL(n)} \Phi_0((Zg, Z))|det\ gZ|^s |det\ Z|^s d^*Z\omega(g)dg\ .$$

If we make the change of variables $Z_1 = ZG$ and $Z_2 = Z$ then this becomes

$$L(s) = \int_{GL(n)} \int_{GL(n)} \Phi_0(Z_1, Z_2)|det\ Z_1|^s |det\ Z_2|^s \omega(Z_2^{-1}Z_1)d^*Z_1 d^*Z_2\ .$$

For $k \in K$, set

$$I(s;k) = \int_{GL(n)} \int_{GL(n)} \Phi_0(Z_1, Z_2)|det\ Z_1|^s |det Z_2|^s \omega(Z_2^{-1}kZ_1)d^*Z_1 d^*Z_2\ .$$

Then $I(s,1) = L(s)$. Since $\Phi_0(k^{-1}Z_1, Z_2) = \Phi_0(Z_1, Z_2)$ and $|det\ k^{-1}Z_1| = |det\ Z_1|$, then by the simple change of variables $Z_1 \to k^{-1}Z_1$ we see that $I(s;k) = I(s;1) = L(s)$ for all $k \in K$. So if dk is the invariant measure on K of total mass 1 we have

$$L(s) = \int_K I(s;k)dk$$
$$= \int_{GL(n)} \int_{GL(n)} \Phi_0(Z_1, Z_2)|det\ Z_1|^s |det\ Z_2|^s \int_K \omega(Z_2^{-1}kZ_1)dk d^*Z_1 d^*Z_2\ .$$

Since v_0 and $\tilde{v}_0$ are K-fixed and $< v_0, \tilde{v}_0 >= 1$, $\omega(g)$ is a zonal spherical function on $GL(n)$. (See, for example, Macdonald [M] for zonal spherical functions.) Therefore we have

$$\int_K \omega(Z_2^{-1}kZ_1)dk = \omega(Z_2^{-1})\omega(Z_1)\ .$$

Since the function $\Phi_0(Z_1, Z_2) = \Phi_0(Z_1)\Phi_0(Z_2)$, with Φ_0 also the characteristic function of $M(n;\sigma)$, we see that the above expression for $L(s)$ splits into separate integrals over Z_1 and Z_2. Since $\omega(Z_2^{-1}) = \tilde{\omega}(Z_2)$, these separate integrations give

$$L(s) = Z(\Phi_0, s, \omega)Z(\Phi_0, s, \omega)$$

and we are done.

6.2. We now turn to the computation of certain local factors for $Sp(2n)$ and the split $O(2n)$ constructed using the local versions of the special Eisenstein series of section 5.

So let $G = O(2n)$ or $Sp(2n)$ and $H = O(4n)$ or $Sp(4n)$, respectively. We will retain the notations from sections 2 and 5. (Note that the dimension of H is multiplied by a factor of 2 from that in section 5.) We fix k to be a non-archimedean local field of characteristic zero.

G acts in the vector space V. Since G is split over k, V admits maximal isotropic subspaces. So let $V = X \oplus Y$ be a polarization of V, i.e., X and Y are totally isotropic and the form on V pairs X and Y non-degenerately. Let $Q \subset G$ be the parabolic subgroup stabilizing Y. Then Q has a decomposition $Q = MU$ with U its unipotent radical and M its Levi subgroup. Then $M \simeq GL(n)$. Let τ be an irreducible unramified representation of M and let π be the irreducible unramified representation of G such that $\pi \subset Ind_Q^G(\tau)$. Let v_0 be a K_M fixed vector in τ (K_M the usual maximal compact of M) and $\tilde{v}_0$ a K_M fixed vector of $\tilde{\tau}$ such that $< v_0, \tilde{v}_0 >_\tau = 1$. Let $\varphi_0 \in \pi$ and $\tilde{\varphi}_0 \in \tilde{\pi}$ be the K fixed vectors such that $\varphi_0(1) = v_0$ and $\tilde{\varphi}_0(1) = \tilde{v}_0$.

The local version of the normalized Eisenstein series $E_H(h;s)$ of section 5 is to take the function

$$f_0(h;s) = d_H(s)\Phi_{K_H,s}(h)$$

as our choice of an element of $ind_P^H(\delta_P^s)$. To simplify notation, we will let $K' = K_H$. Then the local factor we wish to compute is

$$L(s, \varphi_0, \tilde{\varphi}_0, f_0) = d_H(s) \int_G \Phi_{K',s}(g,1) < \pi(g)\varphi_0, \tilde{\varphi}_0 > dg \tag{6.1.}$$

For brevity of notation, we will let $L(s) = L(s; \varphi_0, \tilde{\varphi}_0, f_0)$ for the remainder of this subsection.

We wish to compute $L(s)$ by reducing the calculation to the calculation of a local factor for the representation τ of M. The first step of the reduction is to reduce the integral in (6.1) to an integral over M. If we view elements of π and $\tilde{\pi}$ as elements of $Ind_Q^G(\tau)$ then the pairing of π and $\tilde{\pi}$ can be reduced to the pairing of τ and $\tilde{\tau}$, namely

$$< \varphi, \tilde{\varphi} > = \int_K < \varphi(k), \tilde{\varphi}(k) >_\tau dk$$

where $<,>_\tau$ is the pairing of τ and $\tilde{\tau}$. If we substitute this expression into (6.1) and use the fact that $\tilde{\varphi}_0$ is K-invariant we have

$$\begin{aligned} L(s) &= d_H(s) \int_G \Phi_{K',s}(g,1) \int_K < \varphi_0(kg), \tilde{\varphi}_0(k) >_\tau dk dg \\ &= d_H(s) \int_G \int_K \Phi_{K',s}(k^{-1}g,1) < \varphi_0(g), \tilde{\varphi}_0(1) >_\tau dk dg. \end{aligned} \tag{6.2}$$

We next utilize the following lemma which will be proven in section 6.4 when we analyze $\Phi_{K',s}(g,1)$ as a function on G.

Lemma 6.1. As a function on G, $\Phi_{K',s}(g,1)$ is bi-K-invariant, i.e., for $k_1, k_2 \in K$, $\Phi_{K',s}(k_1 g k_2, 1) = \Phi_{K',s}(g,1)$.

Applying this lemma and integrating over K then gives

$$L(s) = d_H(s) \int_G \Phi_{K',s}(g,1) < \varphi_0(g), \varphi_0(1) >_\tau dg.$$

Next, write $G = QK = UMK$, so that $g = qk = umk$ and $dg = dqdk = \delta_Q(m)^{-1} dudmdk$. Since $\Phi_{K',s}$ and φ_0 are both K invariant we may integrate out K to obtain

$$L(s) = d_H(s) \int_Q \Phi_{K',s}(q,1) < \varphi_0(q), \tilde{\varphi}_0(1) >_\tau dq .$$

Since $\varphi_0 \in Ind_Q^G(\tau)$ we have

$$\varphi_0(q) = \delta_Q^{1/2}(q)\tau(q)\varphi_0(1) = \delta_Q^{1/2}(m)\tau(m)\varphi_0(1)$$

if $q = um$. Recalling that $\varphi_0(1) = v_0$ and $\tilde{\varphi}_0(1) = \tilde{v}_0$, the matrix element in $L(s)$ becomes

$$< \varphi_0(q), \tilde{\varphi}_0(1) >_\tau = \delta_Q^{1/2}(m) < \tau(m)v_0, \tilde{v}_0 >_\tau .$$

As in section 6.1, $< \tau(m)v_0, \tilde{v}_0 >_\tau = \omega_\tau(m)$ is the zonal spherical function on M associated to τ . Hence we may write

$$L(s) = d_H(s) \int_M \left(\int_U \Phi_{K',s}(um,1) du \right) \omega_\tau(m) \delta^{-1/2}(m) dm .$$

The integral over U can be evaluated using the techniques employed in section 5. To this end we write

$$I(h;s) = \int_U \Phi_{K',s}((u,1)h) du .$$

We decompose V^d as $V^d = Y^d \oplus X^d$. As a compliment to V^d in W we may choose $V^d = X^- \oplus Y^+$. Then $W = V_d \oplus V^d = X^- \oplus Y^+ \oplus Y^d \oplus X^d$. With an appropriate choice of bases the form on W is represented in block form by the matrix

$$s = \begin{pmatrix} 0 & 0 & 1 & 0 \\ 0 & 0 & 0 & 1 \\ \varepsilon 1 & 0 & 0 & 0 \\ 0 & \varepsilon 1 & 0 & 0 \end{pmatrix}$$

where $\varepsilon = 1$ for $H = O(2n)$ and $\varepsilon = -1$ for $H = Sp(2n)$. Then, using the notation of Section 5, we have the following lemma.

Lemma 6.2. Let $w \in \Omega$ be the element specified by

$$w^{-1}: \begin{cases} x_1 & \to & x_1 \\ & \vdots & \\ x_n & \to & x_n \\ x_{n+1} & \to & -x_{2n} \\ & \vdots & \\ x_{2n} & \to & -x_{n+1} \end{cases}.$$

Then $U \times 1 = wN_w^- w^{-1}$.

Proof: The unipotent radical S of P can be characterized as those elements s' of H which act by

$$s': \begin{cases} v \to v & v \in V^d \\ v \to \alpha_{s'}(v) + v & v \in V_d \end{cases}$$

with $\alpha_{s'} \in Hom(V_d, V^d)$. Similarly, $U \times 1 \subset H$ can be characterized as those elements $u' \in H$ which act by

$$u': \begin{cases} v \to v + \beta_{u'}(v) & v \in X^+ \\ v \to v & v \in Y^+ \\ v \to v & v \in X^- \\ v \to v & v \in Y^- \end{cases}$$

where now $\beta_{u'} \in Hom(X^+, Y^+)$. So if w is any element of Ω such that

$$w^{-1}: \begin{cases} X^- \to X^- \\ Y^+ \to X^d \\ Y^d \to Y^d \\ X^d \to Y^+ \end{cases}$$

then $w^{-1}Uw \subset S$. From the characterization of Ω in lemma 5.1, we see that the element of Ω which achieves this is the one in the statement of the lemma. Since $U \subset N_{H,\alpha}^-$ (N_H being the unipotent radical of the Borel subgroup B_H of H corresponding to the choice of positive roots) we see that in fact $U = \Pi_{\substack{\alpha<0 \\ w\alpha>0}} N_{H,\alpha}$. Recalling that $N_w^- = \Pi_{\substack{\alpha>0 \\ w\alpha<0}} N_{H,\alpha}$ we have $U = wN_w^{-1}w^{-1}$ as claimed.

Using this fact, then we have

$$\begin{aligned} I(h; s) &= \int_{N_w^-} \Phi_{K',s}(wnw^{-1}h)dw \\ &= c_w(s)\Phi_{K',w^{-1}s}(w^{-1}h) \end{aligned}$$

as in section 5.2. Substituting this back into (6.3) we have completed the first step of our reduction, which we record as a lemma.

Lemma 6.3:

$$L(s) = d_H(s)c_w(s)\int_M \Phi_{K',w^{-1}s}(w^{-1}(m,1))\omega_\tau(m)\delta_Q^{-1/2}(m)dm\ .$$

The next step in the reduction is to find an appropriate integral expression for $\Phi_{K',w^{-1}s}(w^{-1}h)$, similar to that used in section 6.1. For this we need to make more explicit our bases for V and W. Choose bases $\{e_i\}$ for X and $\{f_i\}$ for Y with respect to which the form on V is represented by the matrix

$$\begin{pmatrix} 0 & 1_n \\ \varepsilon 1_n & 0 \end{pmatrix}$$

when $V = X \oplus Y$. So $\varepsilon = 1$ for $G = 0(2n)$ and $\varepsilon = -1$ for $G = Sp(2n)$. Then as bases for our various subspaces of W we choose:

$$\begin{array}{ll} X^+ : \{e_i^+\} & Y^- : \{f_+^-\} \\ Y^+ : \{f_i^+\} & X^d : \{-e_i^+ - e_i^-\} \\ X^- : \{-e_i^-\} & Y^d : \{f_i^+ + f_i^-\}\ . \end{array}$$

Whenever we write elements of H in block form, it will be with respect to these bases.

Now let P_1 be the parabolic subgroup of H stabilizing the flag $V^d \supset Y^d \supset \{0\}$. Then $P \supset P_1 \supset B_H$. Every element $p \in P_1$ determines an automorphism $a(p)$ of Y^d. If we identify $V^d/Y^d \simeq X^d$, then it also determines an element $b(p) \in Hom(X^d, X^d)$. Let μ_s be the character of P_1 defined by

$$\mu_s(p) = |det\ a(p)|^{-1}|det\ b(p)|^{s-(n+2\nu-1)}$$

where $\nu = 1$ for $G = Sp(2n)$ and $\nu = 0$ for $G = 0(2n)$.

Lemma 6.4. $\Phi_{K',w^{-1}s} \in ind_{P_1}^H(\mu_s)$.

Proof: Since $P \supset P_1 \supset B_H$ and $\Phi_{K',w^{-1}s} \in Ind_{B_H}^H(w^{-1}\chi_s)$ it is enough to determine the transformation under the torus $T \subset B_H$. If we write $W = X^- \oplus Y^+ \oplus Y^d \oplus X^d$ and let

$$\underline{t} = \begin{bmatrix} t_1 & & & & & \\ & \ddots & & & & \\ & & t_{2n} & & & \\ & & & t_1^{-1} & & \\ & & & & \ddots & \\ & & & & & t_{2n}^{-1} \end{bmatrix} \in T$$

then

$$a(\underline{t}) = \begin{bmatrix} t_1^{-1} & & \\ & \ddots & \\ & & t_n^{-1} \end{bmatrix} \quad \text{and} \quad b(\underline{t}) = \begin{bmatrix} t_{n+1}^{-1} & & \\ & \ddots & \\ & & t_{2n}^{-1} \end{bmatrix} .$$

Recall that $\chi_s = \sum_{i=1}^{2n}(s+i-2n-\nu)x_i$. Then $w^{-1}\chi_s = \sum_{i=1}^{n}(s+i-2n-\nu)x_i + \sum_{i=n+1}^{2n}(i+\nu-n-s-1)x_i$ and $\delta_B^{1/2}w^{-1}\chi_s = \sum_{i=1}^{n} sx_i - \sum_{i=n+1}^{2n}(s-(n+2\nu-1))x_i$. Therefore $\delta_B^{1/2}w^{-1}\chi_s(\underline{t}) = |det\ a(\underline{t})|^{-s}|det\ b(t)|^{s-(n+2\nu-1)}$.

For the rest of this section we will write H in block form according to the decompositon $W = X^+ \oplus X^- \oplus Y^+ \oplus Y^-$. Let P_Y be the parabolic subgroup of H preserving $Y^+ \oplus Y^-$ and let P_+ be the parabolic subgroup stabilizing the flag $Y^+ \oplus Y^- \supset Y^+ \supset \{0\}$. Then after identifying $(Y^+ \oplus Y^-)/Y^+ \simeq Y^-$, every element $p \in P_+$ determines automorphisms $a_+(p) \in Hom(Y^+, Y^+)$ and $b_+(p) \in Hom(Y^-, Y^-)$. Let λ_s be the character of P_+ defined by

$$\lambda_s(p) = |det\ a_+(p)|^{-s}|det\ b_+(p)|^{s-(n+2\nu-1)} .$$

If we then let

$$A = \begin{bmatrix} 0 & 1 & 0 & 0 \\ 0 & 0 & 1 & 0 \\ 0 & 0 & 0 & 1 \\ -1 & 0 & 0 & 0 \end{bmatrix} \begin{bmatrix} 1 & -1 & 0 & 0 \\ 0 & 1 & 0 & 0 \\ 0 & 0 & 1 & 0 \\ 0 & 0 & 1 & 1 \end{bmatrix}$$

then $P_Y = APA^{-1}, P_+ = AP_1A^{-1}$ and $\mu_s = \lambda_s \circ AdA$.

We may write the elements of $ind_{P_+}^H(\lambda_s)$ in the following form. Let $\Phi_1 \in S(M(2n,4n;k))$ and $\Phi_2 \in S(M(n,4n;k))$. Then define

$$F_1(\Phi_1, s, h) = \int_{GL(2n)} \Phi_1((0,Z)h)|det\ Z|^s d^*Z$$

$$F_2(\Phi_2, s, h) = \int_{GL(n)} \Phi_2((0,0,Z',0)h)|det\ Z'|^s d^*Z'$$

$$F_3(\Phi_2, s, h) = \int_{GL(n)} \Phi_2((0,0,Z',Z')h)|det\ Z'|^s d^*Z' .$$

If Φ_1^0 is the characteristic function of $M(2n,4n;\sigma)$ we let $F_1^0(s,h) = F_1(\Phi_1^0, s, h)$. Similarly for $F_2^0(s,h)$ and $F_3^0(s,h)$. If we then set

$$s_1 = -s + (n + 2\nu - 1)$$

$$s_2 = 2s - (n + 2\nu - 1)$$

then we have, for arbitrary Φ_1 and Φ_2 as above, that

$$F_1(\Phi_1, s_1, h)F_2(\Phi_2, s_2, h) \in ind_{P_+}^H(\lambda_s)$$

$$F_1(\Phi_1, s_1, Ah)F_2(\Phi_2, s_2, Ah) \in ind_{P_1}^H(\mu_s)$$

Lemma 6.5: Let $\xi(s) = \Pi_{i=1}^{2n}\varsigma(-s + n + 2\nu - i)\Pi_{i=1}^{n}\varsigma(2s - n - 2\nu + 2 - i)$ then

$$\Phi_{K',w^{-1}s}(h) = \xi(s)^{-1}F_1^0(s_1, Ah)F_2^0(s_2, Ah) .$$

Proof. We know that $\Phi_{K',w^{-1}s}$ is the unique normalized K_H-fixed vector in $ind_P^H(\mu_s)$. However, as we noted above, $F_1^0(s_1, Ah)F_2^0(s_2, Ah) \in ind_P^H(\mu_s)$ and taking $\Phi_1 = \Phi_1^0$ and $\Phi_2 = \Phi_2^0$ we have guaranteed that this is K'-invariant. So $\Phi_{K',w^{-1}s}$ and $F_1^0(s_1, Ah)F_2^0(s_2, Ah)$ differ only by a function of s. Since $\Phi_{K',w^{-1}s}(1) = 1$ we see that if we set

$$\xi(s) = F_1^0(s_1, A)F_2^0(s_2, A)$$

Then $\Phi_{K',w^{-1}s} = \xi(s)^{-1}F_1^0(s_1, A)F_2^0(s_2, A)$.

To compute $\xi(s)$ we notice that $A \in GL(4n, \sigma)$ so that $F_1^0(s_1, A) = F_1^0(s_1, 1)$ and $F_2^0(s_2, A) = F_2^0(s_2, 1)$. From the identity

$$\int_{GL(m)} \Phi_0(g)|det\ g|^s dg = \prod_{i=1}^{m} \xi(s + 1 + i)$$

we obtain

$$F_1^0(s_1, 1) = \prod_{i=1}^{2n} \varsigma(-s + n + 2\nu - i)$$

and

$$F_2^0(s_2, 1) = \prod_{i=1}^{n} \varsigma(2s - n - 2\nu + 2 - i) .$$

Of course, the function we are really interested in is not $\Phi_{K',w^{-1}s}(h)$ but $\Phi_{K',w^{-1}s}(w^{-1}h)$. Thus we need the next lemma.

Lemma 6.6: $\Phi_{K',w^{-1}s}(w^{-1}h) = \xi(s)^{-1}F_1^0(s_1, h)F_3^0(s_2, h)$.

Proof: In terms of the decomposition $W = X^+ \oplus X^- \oplus Y^+ \oplus Y^-$ the matrix for $w = w^{-1}$ is seen to be

$$w = \begin{bmatrix} 0 & 1 & -v & 0 \\ 0 & 1 & 0 & 0 \\ -v & v & 0 & 0 \\ v & -v & 1 & 1 \end{bmatrix}$$

where

$$v = -\begin{bmatrix} 0 & & 1 \\ & \cdot^{\cdot^{\cdot}} & \\ 1 & & 0 \end{bmatrix} .$$

Then we have

$$Aw = \begin{bmatrix} 0 & 1 & 0 & 0 \\ -v & v & 0 & 0 \\ 0 & 0 & 1 & 1 \\ 0 & 0 & v & 0 \end{bmatrix} .$$

Therefore we see

$$F_1^0(s_1, Awh) = F_1^0(s_1, h)$$

and

$$F_2^0(s_2, Awh) = F_3^0(s_2, h) .$$

Then this lemma follows from lemma 6.5.

We are now ready to complete our reduction to the $GL(n)$ calculation. Let $H_M \subset H$ be the Levi factor of the parabolic subgroup P_Y preserving $Y^+ \oplus Y^-$. Let $P_M \subset H_M$ be the parabolic subgroup given by $P_M = P_Y \cap H_M$. So P_M is the parabolic subgroup of H_M preserving Y_d . Then $M \times M \hookrightarrow H_M$ and $P_M \cap M \times M = M^d$. With respect to the decomposition $W = X^+ \oplus X^- \oplus Y^+ \oplus Y^-$ we have

$$H_M = \left\{ \underline{h}' = \begin{pmatrix} h'^* & 0 \\ 0 & h \end{pmatrix} | h \in GL(2n, k) \right\}$$

and

$$M \times 1 = \left\{ (m, 1) = \begin{bmatrix} m^* & & & \\ & 1 & & \\ & & m & \\ & & & 1 \end{bmatrix} | m \in GL(n, k) \right\} .$$

Here, $h'^* =^t (h')^{-1}$ and $m^* =^t m^{-1}$.

If we consider the restriction of $\Phi_{K',w^{-1}s}(w^{-1}h)$ to the subgroup H_M of H we obtain the following lemma.

<u>Lemma 6.7:</u> For $\underline{h}' \in H_M$ as above,

$$\Phi_{K',w^{-1}s}(w^{-1}\underline{h}') = \xi_1^{-1}(s)|det\ h'|^{s-(n+2\nu-1)} \int_{GL(n)} \Phi_0((Z, Z)h')|det\ h'|^{2s-(n+2\nu_1)} d^*Z$$

where Φ_0 is the characteristic function of $M(n, 2n; \sigma)$ and

$$\xi_1(s) = \prod_{i=1}^{n} \varsigma(2s - n - 2\nu + 2i) \ .$$

Proof: One can easily compute that

$$F_1^0(s_1, \underline{h}') = |det\ h'|^{-s} \prod_{i=1}^{2n} \varsigma(-s + n + 2\nu - i)$$

and

$$F_2^0(s_2, \underline{h}') = \int_{GL(n)} \Phi_2^0((0,0,Z,Z)h')|det\ Z|^2 2d^*Z \ .$$

This and Lemma 6.6 give the formula of the lemma.

If we use this formula in the expression for $L(s)$ given in lemma 6.3 we have

$$L(s) = \frac{d_H(s)c_w(s)}{\xi_1(s)} \int_M |det\ m|^{s-(n+2\nu-1)}$$
$$\int_{GL(n)} \Phi_0((Zm, Z))|det\ Z|^{2s-(n+2\nu-1)} d^*Z \cdot \omega_\tau(m)\delta_Q(m)^{-1/2} dm \ .$$

Easy calculations show that

$$\frac{d_H(s)c_w(s)}{\xi_1(s)} = \begin{cases} 1 & \mathrm{G} = 0(2\mathrm{n}) \\ \varsigma(s-n) & \mathrm{G} = \mathrm{Sp}(2\mathrm{n}) \end{cases}$$

and that due to the way we have written m ,

$$\delta_Q(m) = |det\ m|^{-(n+2\nu-1)} \ .$$

If we place these into the formula above we have, setting $s' = s - \frac{1}{2}(n + 2\nu - 1)$,

$$L(s) = \xi_G(s) \int_M (|det\ m|^{s'} \int_{GL(n)} \Phi_0((Zm, Z))|det\ Z|^{2s'} d^*Z)\omega_\tau(m) dm$$

where $\xi_G(s) = \begin{cases} 1 & \mathrm{G{=}0(2n)} \\ \varsigma(s-n) & \mathrm{G{=}Sp(2n)} \end{cases}$. But this is exactly the $GL(n)$ local factor evaluated in proposition 6.1. Hence we have the following proposition.

Proposition 6.2: For $\pi = Ind_Q^G(\tau)$ with τ an unramified representation of $GL(n)$, φ_0 and $\tilde{\varphi}_0$ the unramified vectors in π and $\tilde{\pi}$, and $f_0(h,s) = d_H(s)\Phi_{K',s}(h)$ we have

$$L(s;\varphi_0,\tilde{\varphi}_0,f_0) = \xi_G(s)Z(\Phi_0,s-\frac{1}{2}(n+2\nu-1),\omega_\tau)Z(\Phi_0,s-\frac{1}{2}(n+2\nu-1),\tilde{\omega}_\tau) .$$

Even when π is not induced from an unramified representation of $GL(n)$, there is a reduction to a combination of $GL(m)$'s and L-functions associated to smaller orthogonal or symplectic groups by this method. Admittedly it is somewhat more complicated. But this reduction allows one to prove the following proposition, which we need in the next section.

Proposition 6.3: Let k be a local field. Let π be an irreducible admissible representation of G , $\varphi_1 \in \pi$, and $\varphi_2 \in \tilde{\pi}$. Let $f \in ind_P^H(\delta_P^s)$ and assume that $f(h;s)$ is holomorphic in s for fixed $h \in H$.

(i) If k is non-archimedean, then the local integral $L(s;\varphi_1,\varphi_2,f)$ is a rational function in q^s and q^{-s}, q being the cardinality of the residue field. Moreover, the denominator can be chosen independent of φ_1,φ_2 and f ; it only depends on the representation π .

(ii) If k is archimedean, then the local integral $L(s;\varphi_1,\varphi_2,f)$ is a polynomial in s times a product of Γ-functions.

6.3. We now come to the main result of this section. Take k to be a number field and G to be either $Sp(2n)$ or an $SO(2n)$, split over k . The connected component of the L-group of G is then

$$L_G0 = \begin{cases} SO(2n+1,\ \mathbb{C}) & G = Sp(2n) \\ SO(2n,\ \mathbb{C}) & G = SO(2n) \end{cases} .$$

Since G is split over k, the full group ${}^LG$ is simply the product ${}^LG = {}^LG^0 \times W_k$, where W_k is the Weil group of k . Let r be the standard representation of ${}^LG$. r is trivial on W_k and on ${}^LG^0 = SO(m,\ \mathbb{C})$ it is given by the embedding $SO(m,\ \mathbb{C}) \hookrightarrow GL(m,\ \mathbb{C})$.

Let $\pi = \otimes\pi_v$ be an irreducible cuspidal automorphism representation of $G_{\mathbf{A}}$. Let $S = S(\pi)$ be the set of all finite places of k at which π_v is unramified. This is all but a finite number of all places of k. Then to π and r Langlands has associated an L-function

$L(s,\pi,r)$ given by a product over the places in S of local factors

$$L(s,\pi,r) = \prod_{v \in S} L_v(s,\pi_v,r_v) .$$

(See Borel's article in Corvallis [B].) Let B_v be a Borel subgroupof G_v and π_v be the unique component of $Ind_{B_v}^{G_v}(\mu_1,\cdots,\mu_n)$ which contains a K_v fixed vector (call this the unramified component). Here induction is taken in a normalized sense and $\mu_1,\cdots,\mu_n$ are unramified characters on the split torus $T_v \subset B_v$. Then these local factors are given by

$$L_v(s,\pi_v,r) = \begin{cases} \prod_{i=1}^n L_v(s,\mu_i)L_v(s,\mu_i^{-1}) & G = SO(2n) \\ \zeta_v(s)\prod_{i=1}^n L_v(s,\mu_i)L_v(s,\mu_i^{-1}) & G = Sp(2n) \end{cases}$$

where the $L(s,\mu_i)$ are the Tate L-factors .

Theorem 6.1.: Let $G = Sp(2n)$ or the split $SO(2n)$. Let π be an irreducible cuspidal automorphic representation of $G_{\mathbb{A}}$ and r the standard representation of ${}^LG$. Then the Langlands L-functions $L(s,\pi,r)$ have a meromorphic continuation to $\mathbb{C}$ with only a finite number of poles.

Proof: We will compare the Langlands L-function to the L-function of section 1 constructed using the normalized Eisenstein series of section 5. However, in the case of the orthogonal groups, Langlands' construction is for $SO(2n)$ while the construction of section 1 is for $O(2n)$. So we must first pass from $SO(2n)$ to $O(2n)$.

Assume now that $G = SO(2n)$ and let $G' = O(2n)$ so that $G' \supset G$. Then by a standard construction there exists a cuspidal representation π' of $G'(\mathbb{A})$which contains π as a summand by restriction to $G(\mathbb{A})$.

At almost all places $v \in S(\pi)$ the representation π' will also be unramified, and infact, if π_v is the unramified component of $Ind_{B_v}^{G_v}(\mu_{1,v}\cdots\mu_{n,v})$ then $(\pi')_v$ is the unramified component of $Ind_{B'_v}^{G_v}(\mu_{1,v}\cdots\mu_{n,v})$. We let $S' \subset S$ be this set of places.

For $G = Sp(2n)$ we set $G' = G$, $\pi' = \pi$, and $S' = S$ so that we may treat $Sp(2n)$ and $O(2n)$ simultaneously.

Let

$$L_{S'}(s,\pi,r) = \prod_{v \in S'} L_v(s,\pi_v,r_v) .$$

Then since S and S' differ in only a finite number of places, $L(s,\pi,r)$ and $L_{S'}(s,\pi,r)$ will differ only by a finite number of Euler factors. Hence it is enough to prove the theorem for $L_{S'}(s,\pi,r)$.

Let H be either $Sp(4n)$ or $0(4n)$, constructed as in section 2 from G'. Take $\varphi \in \pi'$ and $\tilde{\varphi} \in \tilde{\pi}'$ decomposable such that φ_v and $\tilde{\varphi}_v$ are unramified for $v \in S'$ and $< \varphi_v, \tilde{\varphi}_v >= 1$. For each finite place $v \notin S'$ we can choose a function $\beta_v \in \mathcal{H}_v$, the local Hecke algebra on H_v, such that for $f_v = \Phi_{K_v,s} * \beta \in ind_{P_v}^{H_v}(\delta_P^s)$ we have

$$\int_{G_v} f_v(g,1) < \pi'_v(g)\varphi_v, \tilde{\varphi}_v > dg = 1 \ .$$

This choice will depend on φ_v and $\tilde{\varphi}_v$, but that will not matter for now. Choosing such a β_v for each finite $v \notin S'$, let $\beta = \otimes\beta_v \in \prod_{\substack{v<\infty \\ v\notin S'}} \mathcal{H}_v$ considered as an element in $\mathcal{H}_{\mathbb{A}}$. For such a β we have constructed an Eisenstein series $E_H(h;s;\beta)$ in section 5 and by Theorem 5.1 we know that $E_H(h;s;\beta)$ has a meromorphic continuation with only a finite number of poles.

Consider now the global L-integral

$$L(s;\varphi,\tilde{\varphi},f) = \int_{(G'\times G')_k\backslash(G'\times G')_{\mathbb{A}}} E_H((g_1,g_2);s;\beta)\varphi(g_1)\tilde{\varphi}(g_2)dg_1dg_2$$

where $f(h,s) = d_H(s)\Phi_{K,s} * \beta$. Then from the analytic properties of $E_H(h;s;\beta)$ we know that $L(s;\varphi,\tilde{\varphi},f)$ has a meromorphic continuation in S with only a finite number of poles. On the other hand, this integral also factors into the product of local integrals

$$L(s;\varphi,\tilde{\varphi},f) = \prod_v L_v(s;\varphi_v,\tilde{\varphi}_v,f_v) \ .$$

At the places $v \in S'$, $f_v = d_{H_v}(s)\Phi_{K_v,s}$ and these local integrals were evaluated in section 6.2. There we found

$$L_v(s;\varphi_v\tilde{\varphi}_v,fc) = \xi_{G'}(s)Z(\Phi_0,s',\omega_\tau)Z(\Phi_0,s',\tilde{\omega}_\tau)$$

where $s' = s-\frac{1}{2}(n+\nu-1)$ and $\tau = Ind_{B'_v\cap M_v}^{M_v}(\mu_{1,v},\cdots,\mu_{n,v})$ if $\pi'_v = Ind_{B'_v}^{G'_v}(\mu_{1,v},\cdots,\mu_{n,v})$. But by [G-J] we know that

$$Z(\Phi_0,s',\omega_\tau) = \prod_{i=1}^{n} L_v(s-n+1-\nu,\mu_{i,v})$$

$$Z(\Phi_0,s',\tilde{\omega}_\tau) = \prod_{i=1}^{n} L_v(s-n+1-\nu,\mu_{i,v}^{-1}).$$

Therefore, for $v \in S'$, we have

$$L_v(s;\varphi_v,\tilde{\varphi}_v,f_v) = L_v(s,\pi_v,r_v) \ .$$

If v is a finite place, but $v \notin S'$, then $f_v = d_{H,v}(s)\Phi_{K_v,s} * \beta_v$. By our choice of β_v we have

$$L_v(s;\varphi_v,\tilde{\varphi}_v,f_v) = d_{H,v}(s) \ .$$

If v is an archimedean place, then by proposition 6.3 we have that $L_v(s;\varphi_v,\tilde{\varphi}_v,f_v)$ is $d_{H,v}(s)$ times a polynomial in s and a product of Γ-functions. Therefore we see that the finite product

$$\prod_{v \notin S'} L_v(s;\varphi_v\tilde{\varphi}_v,f_v)$$

is a meromorphic function of s with only a finite number of zeroes.

Combining these local calculations we have

$$L(s,\varphi,\tilde{\varphi},f) = L_{S'}(s,\pi,r)(\prod_{v \notin S'} L_v(s;\varphi_v,\tilde{\varphi}_v,f_v)) \ .$$

Since the left hand side is meromorphic in s with only a finite number of poles and the finite product over $v \notin S'$ is meromorphic in s with a finite number of zeroes, we may conclude that $L_{S'}(s,\pi,r)$ is meromorphic in s with only a finite number of poles. This proves the theorem.

By calculations similar to those in the above theorem we can show that for any special orthogonal group $SO(n)$ the corresponding Langlands L-function $L(s,\pi,r)$ (for the standard representation r) has a meromorphic continuation in s admitting only a finite number of poles. In particular, this covers the case when $SO(n)$ is anisotropic, where it is not possible to use Whittaker model techniques.

6.4. In the computation of the local factors for $O(2n)$ and $Sp(2n)$ we have used the fact that the function $\Phi_{K',s}(g,1)$ was bi-K-invariant. (K' is the maximal compact in H and K that of G.) In this section we will prove this fact along the way to deriving an explicit formula for $\Phi_{K',s}(g,1)$. We return the basic notation of section 6.2.

The computation is a local one, so we let k be a local field, archimedean or not. We will first derive an integral representation for $\Phi_{K',s}(h)$. When k is non-archimedean, we set Φ_0 to be the characteristic function of $M(2n,4n;\sigma)$. For k archimedean, we see

$$\Phi_0(x) = \begin{cases} exp(-tr({}^t xx)) & k = \mathbb{R} \\ exp(-tr({}^t \bar{x}x)) & k = \mathbb{C} \end{cases}$$

for $x \in M(2n,4n;k)$. In either case, $\Phi_0 \in S(M(2n,4n;k))$ and it is right invariant under K_{4n}, the maximal compact of $GL(4n,k)$, and is left invariant under K_{2n}, the maximal compact of $GL(2n,k)$.

We write elements of H or $GL(4n,k)$ in block form according to the decompositions $W = V^+ \oplus V^- = X^+ \oplus Y^+ \oplus X^- \oplus Y^-$. If we let

$$A = \begin{bmatrix} 0 & 1 & 0 & 0 \\ 0 & 0 & 1 & 0 \\ 1 & 0 & 1 & 0 \\ 0 & 1 & 0 & 1 \end{bmatrix}$$

then A conjugates the parabolic P in H which stabilizes V^d into the parabolic P' of $GL(4n,k)$ having block form

$$P' = \left\{ P' = \begin{pmatrix} * & * \\ 0 & a \end{pmatrix} \right\} .$$

Then we have, in this form, $\delta^s_{P'}(P') = |det\ a|^{-s}$. If we then construct

$$F'_0(h,s) = \int_{GL(2n)} \Phi_0((0,Z)Ah)|det\ Z|^s d^*Z$$

we see that $F' \in ind_P^H(\delta_P^s)$ and is K'-invariant. If we then normalize by setting $F_0(h,s) = \xi(s)^{-1} F'_0(h,s)$ where

$$\xi(s) = F'_0(1,s) = \int_{GL(n)} \Phi_0(0,Z)|det\ Z|^s d^*Z$$

then $\Phi_{K',s}(h) = F_0(h,s)$. If we restrict this to the group

$$G \times 1 = \left\{ \begin{pmatrix} g & 0 \\ 0 & 1 \end{pmatrix} | g \in G \right\}$$

we have

$$\begin{aligned} \Phi_{K',s}(g,1) &= \xi(s)^{-1} \int_{GL(2n)} \Phi_0(Zg,Z)|det\ Z|^s d^*Z \\ &= \xi(s)^{-1} \int_{GL(2n)} \Phi_0(Z(g,1))|det\ Z|^s d^*Z . \end{aligned}$$

From this last formula, the bi-K-invariance of $\Phi_{K',s}(g,1)$ is clear and we have proven lemma 6.1, even in the archimedean case.

For any $A \in M(2n,4n;k)$, let

$$\kappa(A) = \xi(s)^{-1} \int_{GL(2n)} \Phi_0(ZA)|det\ Z|^s d^* Z .$$

Then $\kappa(A)$ is a function on $M(2n,4n;k)$ which is right invariant under K_{4n} , left invariant under K_{2n} , and positive homogeneous of degree $-2ns$. On the other hand we have the following elementary fact.

Lemma 6.8: Up to scalar multiples, there is a unique function on $M(2n,4,k)$ which satisfies

a) it is right invariant under K_{4n}

b) it is left invariant under K_{2n}

c) it is positive homogeneous of degree $2n$.

This function is the 2n-dimensional volume of the parallelipiped in k^{4n} spanned by the rows of the matrix $A \in M(2n,4n,k)$.

We will denote this function by vol (A). Then it is clear that we have $\kappa(A) = (vol A)^{-s}$. Since $\Phi_{K',s}(g,1) = \kappa(g,1)$, we will be able to explicitely evaluate $\Phi_{K',s}$ by finding a simple formula for the volume function.

For $A \in M(2n,4n;h)$, let $\mathcal{M}(A)$ be the set of $2n \times 2n$ minors of A. If k is non-archimedean, set

$$\kappa_2(A) = \max_{B\in \mathcal{M}(A)} (|det\ B|) .$$

If k is archimedean, set

$$\kappa_2(A) = (\sum_{B\in \mathcal{M}(A)} |det\ B|^2)^{1/2} .$$

Then it is easy to check that κ_2 satisfies the properties a), b) and c) of lemma 6.8 and that in fact $\kappa_2(A) = vol(A)$.

To restrict κ_2 to matrices of the form $A = (g,1)$ we set $\mathcal{M}_1(g)$ to be the set of all minors of g , with 1 taken to be the minor of degree 0. Then an elementary calculation gives

$$\kappa_2(g,1) = \max_{a\in \mathcal{M}_1(g)} |det\ a|$$

for k non-archimedean and

$$\kappa_2(g,1) = (\sum_{a\in \mathcal{M}_1(g)} |det\ a|^2)^{1/2} .$$

Now let k be non-archimedean. By elementary divisor theory we may write any $g \in G$ as $g = k_1 d k_2$ with $k_1, k_2 \in K$ and $d = diag(\omega^{a_1}, \cdots, \omega^{a_n}, \omega^{-a_1}, \cdots, \omega^{-a_n})$ with $a_1 \geq a_2 \geq \cdots \geq a_n \geq 0$, where ω is a uniformizer for σ. Then we have

$$\kappa_2(g,1) = \kappa_2(d,1) = q^{\Sigma_{i=1}^n a_i}$$

and so

$$\Phi_{K',s}(g,1) = \kappa_2(g,1)^{-s} = q^{-s\Sigma_{i=1}^n a_i} .$$

If k is archimedean, then we can use the Cartan decomposition to write $g = k_1 d k_2$ with $k_1, k_2 \in K$ and $d = diag(d_1, \cdots, d_n,\ d_1^{-1}, \cdots, d_n^{-1})$ with the $d_i \in \mathbb{R}$ and all $d_i > 0$. Then we have

$$\kappa_2(g,1) = \kappa_2(d,1) = (\prod_{i=1}^{n} (1 + d_i^2)(1 + d_i^{-2}))^{1/2}$$

so that

$$\Phi_{K',s}(g,1) = \kappa_2(g,1)^{-s} = \left[\prod_{i=1}^{n} (1 + d_i^2)(1 + d_i^{-2}) \right]^{-s/2} .$$

We record these facts in this last proposition.

Proposition 6.4. a) Let k be a non-archimedean and write $g \in G$ as $g = k_1 d k_2$ with $k_1, k_2 \in K$ and $d = diag(\omega^{a_1}, \cdots, \omega^{a_n},\ \omega^{-a_1}, \cdots, \omega^{-a_n})$ with $a_1 \geq a_2 \geq \cdots \geq a_n \geq 0$. Then

$$\Phi_{K',s}(g,1) = q^{-s\Sigma_{i=1}^n a_i} .$$

b) Let k be archimedean and write $g \in G$ as $g = k_1 d k_2$ with $k_1, k_2 \in K$ and $d = diag(d_1, \cdots, d_n,\ d_1^{-1}, \cdots, d_n^{-1})$ with $d_i \in \mathbb{R}$ and $d_i > 0$. Then

$$\Phi_{K',s}(g,1) = \left[\prod_{i=1}^{n} (1 + d_i^2)(1 + d_i^{-2}) \right]^{-s/2} .$$

References

[A] J. Arthur, Eisenstein series and the trace formula. Pro. Symp. Pure Math. 33 (1979), pt. 1, 253 - 274.

[B] A. Borel, Automorphic L-functions. Pro. Symp. Pure Math. 33 (1979), pt. 2, 27 - 61.

[C] W. Casselman, Introduction to the theory of admissible representations of p-adic reductive groups (manuscript).

[G-J] R. Godement and H. Jacquet, Zeta Functions of Simple Algebras. Lecture Notes in Math., vol. 260, Springer, New York.

[G-K] S. Gindikin and F. Karpelevich, On an integral connected with symmetric Riemann spaces of nonpositive curvature. Amer. Math. Soc. Transl. (2) 85 (1969), 249 - 258.

[L1] R. Langlands, Euler Products. Yale University Press, 1967.

[L2] R, Langlands, On the functional equation satisfied by Eisenstein series. Lecture Notes in Math., vol 544, Springer, New York.

[M] I. Macdonald, Spherical functions on a group of p-adic type. Ramanujan Institute, University of Madras Publ., 1971.

PART B : L-functions for $G \times GL(n)$

by

S. Gelbart and I. Piatetski-Shapiro

Introduction.

Let G' denote a connected reductive algebraic group defined over a global field k, and Π an automorphic cuspidal representation of $G'(\mathbb{A}_k)$. To each holomorphic finite-dimensional representation r of the Langlands dual group ${}^LG'$, one may attach an automorphic L-function

$$L(s,\Pi,r) = \Pi_v L(s,\Pi_v,r)$$

with $L(s,\Pi_v,r)$ the "local Langlands factor" at v; cf. Sections 12-14 of [Borel 1] or [Langlands 1]. We recall that this Euler product is defined only over the unramified places of k, with respect to Π, and for these places

$$L(s,\Pi_v,r) = [det[I - r(t_v)(Nv)^{-s}]^{-1},$$

with $\{t_v\}$ the conjugacy class in ${}^LG'$ uniquely determined by Π_v. The most famous example of these L-functions are the "standard automorphic L-functions" attached to $G' = GL(n)$, with r the standard representation of ${}^LG' = GL(n, \mathbb{C})$ (given by the identity map $g \to g$).

The Euler product defining $L(s,\Pi,r)$ is known to converge in some right half-plane ([Langlands 1]). A fundamental problem in the theory of automorphic forms is to prove the following:

Conjecture ([Langlands 1]).

The Euler product defining $L(s,\Pi,r)$ continues to a meromorphic function in the whole s-plane, with only finitely many poles; moreover, $L(s,\Pi,r)$ satisfies a functional equation relating its value at s to its value at $1-s$.

Implicit in this Conjecture is the possibility of defining $L(s,\Pi_v,r)$ at the "ramified" places as well. For the standard automorphic L-functions for $GL(n)$, this Conjecture was established by [Godement- Jacquet] using a simple integral expression for $L(s,\Pi,r)$; in this case $L(s,\Pi,r)$ was shown to have no poles at all. More recently, Piatetski-Shapiro and Rallis have been adapting the ideas of Godement-Jacquet to obtain similar results for G' an arbitrary simple classical group; cf. [PSR1], [PSR2] (and work in progress). The methods of Piatetski-Shapiro and Rallis are actually based on a proper generalization of the classical Rankin-Selberg trick first formulated representation theoretically in [Jacquet]. In addition to embracing almost all the L-functions known classically, the results of Piatetski-Shapiro and Rallis include whole families of new automorphic L-functions, most notably those for the orthogonal and symplectic groups. In all these cases, r is taken to be the standard embedding of ${}^LG'$ in the appropriate general linear group.

In the present paper, we shall describe new generalizations of the Rankin-Selberg method which make it possible to prove Langlands' Conjecture for non-simple groups of the form $G \times GL(n)$, at least when G has split *rank* n. For these L-functions, the proof of the conjecture is ultimately related to the lifting of automorphic forms between G and an appropriate $GL(m)$. A special example of this phenomenon already appears in the work of Piatetski-Shapiro and Soudry on $GSp_4 \times GL_2$, and in the work of the present authors on $U_{2,1} \times GL_1(K)$. In [Soudry], [PSS1] and [PSS2], the relevant lift is from GSp_4 to GL_4; on the other hand, in [GePS] the lifting is a "base change lift" from $U_{2,1}$ to $GL_3(K)$, with K the quadratic extension of k defining the unitary group $U_{2,1}$.

A prototype of our Rankin-Selberg approach may be described as follows. Suppose G is the split special orthogonal group in $2n+1$ variables over k, and H denotes the subgroup of G isomorphic to $SO(2n)$ (the split special orthogonal group in $2n$ variables). Then H has a maximal parabolic subgroup P whose Levi component is isomorphic to $GL(n)$. Given automorphic cuspidal representations π and τ, of G and $GL(n)$respectively, we form the integral

$$I(s,\varphi,E)=\int_{H_k\backslash H_A}\varphi(h)E^{\tau}(h,s)dh$$

where φ belongs to the space of π, and E is an Eisenstein series on H associated to τ, i.e., induced from the cuspidal representation $\tau\otimes|det|^{s'}$ of P, with $s'=s-1/2$. Assuming that both π and τ possess Whittaker models, we can establish a basic identity.

$$(*)\qquad I(s,\varphi,E)=\int_{U_{\mathbb{A}}^H\backslash H_{\mathbb{A}}}W(h)W_{\tau}(h,s)dh$$

where $U_{\mathbb{A}}^H$ is the standard maximal unipotent subgroup of H, $W(h)$ belongs to the Whittaker model of π, and $W_{\tau}(h,s)$ belongs to the space of *ind* $\tau\otimes|det|^{s'}$.

It turns out that the right hand side of (*) factors naturally as a product of similar integrals over each of the places v of k, thereby defining a local zeta-integral attached to the pair of representations π_v,τ_v. For each of these zeta-integrals one can prove a functional equation, and for "unramified" v, one can show that the appropriate zeta-integral coincides with the local Langlands factor $L(s,\Pi_v,r)$. In this way, the analytic continuation and functional equation of $L(s,\Pi,r)$ can be derived from the corresponding properties of the left-hand side of (*), i.e., from the Eisenstein series $E(h,s)$. (Note that Π here refers to a pair of representations π,τ.)

In general, for groups G' of the form $G\times GL(n)$, we are led to consider local factors $L(s,\Pi,r)$, with Π still a pair of representations (π,r). In particular, when $G=GL(n)$, we are back to the theory of Rankin-Selberg convolutions for $GL(n)\times GL(n)$ developed in [JPS S1]. In this paper, we therefore deal primarily with the split groups of type B_n, C_n and D_n. Although our theory can also be adapted to deal with non-split groups such as the unitary group, we shall deal only parenthetically with such groups in the present paper; cf. §6 for further remarks.

In Chapter I, we establish the basic identity for the split classical groups $G=SO_{2n+1}, SO_{2n}$ and Sp_{2n}. Although the form of the integral $I(s,\varphi,E)$ looks similar for each family of orthogonal groups, the actual derivation of the identity requires different methods depending on whether the dimension of the form is even or odd. For the symplectic groups, even the form of the integral for $I(s,\varphi,E)$ is new, requiring as it does a theta-series in the integral. In order to emphasize the underlying simplicity (and generality) of our work, we begin Chapter I with an axiomatic treatment of these three methods. The rest of the Chapter then amounts to a case by case check of these axioms for each of the special examples considered.

In Chapter II, we establish the functional equation for the local zeta-integrals which arise from the factorization of the right-hand side of the basic identity, again by separate arguments for each of the families B_n, C_n and D_n. Implicit here is the meromorphic continuation of these local integrals as rational functions of q^{-s} (q the residual characteristic of k_v). Rather than establishing these rationality results by familiar methods special to each of the families of groups considered, we appeal to a recent elegant approach to the problem due to J. Bernstein.

In Section 13, we discuss the normalization of our Eisenstein series and local zeta-integrals. The resulting local "unramified" zeta-integrals are computed using an approach inspired by S. Rallis; for this, see the Appendix to our paper, written jointly with Rallis.

In forthcoming works, we hope to give a more complete analysis of our local zeta-integrals and L-factors, at the "ramified" places as well. The goal in general will be to define L and ε factors at each place in terms of g.c.d.'s of the local zeta-integrals, and prove "inductivity properties" for these local factors when the Π_v (i.e. Π_v and τ_v) are induced representations. Unfortunately, even the formulation of these factorization results involves an analogous theory of local factors for $G \times GL(m)$ with m less than the split rank of G.

We are grateful to Miriam Abraham for her efficient and patient preparation of the manuscript, and David Soudry for several helpful conversations.

POST SCRIPT

After the final draft of this manuscript was prepared in the Spring of 1986, we spoke with F. Shahidi about his recent work on automorphic L-functions ([Shahidi 2]). By further refining his earlier extension of Eisenstein series methods initiated in [Langlands 2], Shahidi has been able to establish the finiteness of poles for a remarkably general class of automorphic L-functions without using explicit integral expressions like our $I(s, \varphi, E)$.

Shahidi's results do not apply directly to our $L(s, \pi \times \tau)$. However, they do imply that there are only finitely many poles for the $GL(n)$ functions $L(s, \tau, Sym^2)$ and $L(s, \tau, \Lambda^2)$ alluded to in Section 13), and they also apply to the standard L-functions $L(s, \pi)$ of [PSR1], provided the representations in question are generic. A description of Shahidi's theory is to appear soon as a research announcement of the Bulletin of the A.M.S.. Although Shahidi's approach appears to be more general than ours, it seems that our use of integral expressions will ultimately lead us closer to an explicit description of the (possibly finitely many) poles of our L-functions.

Chapter I: Basic Identities and the Euler Product Expansion

§1. Description of Possible Methods

We shall outline here three different methods for introducing global zeta-integrals $I(s,\varphi,E)$ with Euler product expansion. In each case, φ will belong to an irreducible space of cusp forms π on $G_{\mathbb{A}}$, and E will be an Eisenstein series on the reductive group $H_{\mathbb{A}}$. More specifically, E will be parabolically induced from the cuspidal representation τ of $GL(n)$, with $GL(n)$ realized as the Levi component of a parabolic subgroup P of H. For the first method (respectively the second), H will be a subgroup of G (respectively G will be a subgroup of H); for the third method - Method C, H equals G.

As already suggested in the Introduction, we shall carry out detailed applications of these methods only for some special classes of groups. In particular, Method A will be applied to split groups G of type B_n, Method B to groups of type D_n, and Method C to the symplectic groups. It is important to emphasize, however, that each of these methods actually has a wider range of applicability. For example, both methods A and B may be applied to groups G of type B_n, the former yielding L-functions on $G \times GL(n)$, the latter L-functions on $G \times GL(n+1)$. Such results might be useful to future work on the afore-mentioned lifting problem via converse theorems for L-functions. Moreover, all three methods apply to other classes of groups such as the spin groups and quasi-split unitary groups; cf. §6.

Before describing these methods axiomatically, we mention one result from [PS] which underlies them all. Fix a $\mathbb{A}$-field k and a non-trivial additive character ψ of $k \setminus \mathbb{A}$. Let P_n denote the subgroup of matrices

$$\left\{ \begin{bmatrix} & & & * \\ & g & & \vdots \\ & & & * \\ 0 & \cdots & 0 & 1 \end{bmatrix} \; : \; g \in GL(n-1) \right\}$$

in $GL(n)$. By a standard cuspidal subgroup of P_n we mean a subgroup of P_n consisting of all matrices of the form

$$\begin{bmatrix} I_{k_1} & \cdots & * \\ & \ddots & \\ 0 & & I_{k_r} \end{bmatrix}, \quad k_1 + k_2 + \cdots + k_r = n.$$

Consider functions $\alpha^*(p)$ in $C^\infty(P_n(k) \setminus P_n(\mathbb{A}))$ such that α^* is cuspidal, i.e., $\int_{R_k \setminus R_A} \alpha^*(rp)dr = 0$ for every standard cuspidal subgroup R of P_n. Then

(1.0.1)
$$\alpha^*(p) = \sum_{\delta \in Z_n \setminus P_n} \alpha^*_\theta(\delta p),$$

where Z_n is the subgroup of upper triangular unipotent matrices in P_n, $\theta(z) = \psi(\sum_{i=1}^{n-1} z_{i,i+1})$, and

(1.0.2)
$$\alpha^*_\theta(p) = \int_{Z_n(k) \setminus Z_n(\mathbb{A})} \alpha(zp)\overline{\theta(z)}ds.$$

Moreover, the series in (1.0.1) converge absolutely and uniformly. This Fourier expansion plays a crucial role in our proof of the basic identity for $I(s, \varphi, E)$, and it comes into play precisely because our groups G are so closely related to $GL(n)$.

We now proceed to describe our three general types of zeta-integrals and basic identities. The Euler product expansion for these zeta-integrals will be discussed only after the basic identities are obtained.

(1.1.) Method A.

The relevant data here is

$$\begin{matrix} H & \subset & G \\ \cup & & \cup \\ P & \subset & Q \end{matrix}$$

where H and G are reductive groups, P is a parabolic subgroup of H with Levi decomposition

$$P = GL(n)U^P,$$

and Q is a (non necessarily parabolic) subgroup of G satisfying the following properties:

(1.1.1.) U^P is normal in Q, and fits into the exact sequence

$$1 \to U^P \to Q \xrightarrow{\alpha} P_{n+1} \to 1;$$

(1.1.2) $\alpha^{-1}(Z_{n+1})$ defines a maximal unipotent subgroup U^G of G, and $\theta \circ \alpha$ defines a non-degenerate character of U^G.

Assuming this data, let us consider the global zeta integral

$$I(s,\varphi,E) = \int_{H_k \backslash H_{\mathbb{A}}} \varphi(h) E_f^{\tau}(h,s)\,dh \tag{1.1.3}$$

where φ is a cusp form on G (belonging to the irreducible representation π), and

$$E_f^{\tau}(h,s) = \sum_{\gamma \in P_k \backslash H_k} f^{\tau}(\gamma h, s). \tag{1.1.4}$$

Here f^{τ} is in the space of the induced representation $ind_{P_{\mathbb{A}}}^{G_{\mathbb{A}}} \tau \otimes |det|^{s'}$, with $s' = s - 1/2$. Since φ is rapidly decreasing when restricted to H, the integral (1.1.3) is absolutely convergent in the same half-plane where $E_f^{\tau}(h,s)$ converges absolutely and uniformly.

Here we are viewing $f^{\tau}(h,s)$ concretely as a complex-valued function on $H_{\mathbb{A}}$ which is left $U_{\mathbb{A}}^P$ - invariant and such that for each fixed $h \in H_{\mathbb{A}}$, the function $w(m) = f^{\tau}(mh,s)$ belongs to the space of cusp forms on $GL(n,\mathbb{A})$ realizing the representation $\tau \otimes |det(m)|^{s'+\frac{n-1}{2}}$ (the factor $|det(m)|^{\frac{n-1}{2}}$ being the squareroot of the modulus function on P). The isomorphism between this space and the familiar abstract space of functions with values in the space of $\tau \otimes \| \|^{s'}$ such that $f'(ph) = |det\ m|^{s+\frac{n-1}{2}} \tau(p) f'(h)$ is provided by the map $f \to f'(h)(m) = f(mh,s)$.

Theorem A. With date H, G, P, Q satisfying axioms (1.1.1) and (1.1.2), we have

$$I(s,\varphi,E) = \int_{U_{\mathbb{A}}^H \backslash H_{\mathbb{A}}} W_{\varphi}(h) W_f^{\tau}(h,s)\,dh \tag{1.1.5}$$

with $U_{\mathbb{A}}^H = Z_n(\mathbb{A}) U_{\mathbb{A}}^P$,

$$W_{\varphi}(g) = \int_{U_k^G \backslash U_{\mathbb{A}}^G} \varphi(ug) \overline{\theta \circ \alpha}(u)\,du$$

and

$$W_f^\tau(h,s) = \int f^\tau(zh,s)\theta(z)dz.$$

Proof. It follows from the definition of $E_f^\tau(h,s)$ (and the left invariance of f by $U_{\mathbb{A}}^P$) that

$$I(s,\varphi,E) = \int_{P_k\backslash H_{\mathbb{A}}} \varphi(h)f^\tau(h,s)dh = \int_{P_k U_{\mathbb{A}}^P\backslash H_{\mathbb{A}}} \varphi^\circ(h)f(h,s)dh$$

with

$$\varphi^\circ(h) = \int_{U_{\mathbb{A}}^P\backslash U_{\mathbb{A}}^P} \varphi(u'g)du'.$$

Because of the normality of U^P in Q, it is easy to check that $\varphi^\circ(g)$ is left Q_k-invariant. Therefore, by assumption (1.1.1), the formula $\alpha^*(p) = \varphi^\circ(\alpha^{-1}(p)g)$ determines a well-defined cuspidal function in $C^\infty(P_{n+1}(k)\setminus P_{n+1}(\mathbb{A}))$ for any g in $G_{\mathbb{A}}$. In particular, by (1.0.1) we obtain the Fourier expansion

(1.1.6)
$$\varphi^\circ(g) = \alpha^*(1) = \sum_{\delta\in Z_n\backslash GL_n} W_\varphi(\delta g).$$

Thus

$$I(s,\varphi,E) = \int_{U_{\mathbb{A}}^P P_k\backslash H_{\mathbb{A}}} \left\{\sum_{Z_n\backslash GL_n(k)} W_\varphi(\delta h)\right\} f^\tau(h,s)dh.$$

Since $P = GL_nU^P$ implies $Z_n\setminus GL_n(k) \approx Z_n\, U_{\mathbb{A}}^P \setminus P_n U_{\mathbb{A}}^P$,

$$I(s,\varphi,E) = \int_{Z_n U_{\mathbb{A}}^P\backslash H_{\mathbb{A}}} W_\varphi(h)f^\tau(h,s)dh.$$

Integrating with respect to $Z_n(k)\setminus Z_n(\mathbb{A})$ then gives the desired identity (1.1.5).

Concluding remark. The right side of our basic identity will be identically zero unless π and τ possess Whittaker models, i.e. are "non-degenerate". Indeed, the space of functions $\{W_\varphi : \varphi \textit{ in the space of } \pi\}$ comprises the Whittaker model $\mathcal{W}(\pi,\psi)$ for π with respect to the "non-degenerate" character $\psi_U = \theta\circ\alpha$ of U^G; similarly the space of functions $\{W^\tau(mh,s)\}$ belongs to the Whittaker model of τ with respect to θ^{-1}. The Fourier expansion (1.1.6) for φ° implies that $\mathcal{W}(\pi,\psi)\equiv 0)$ if and only if $\varphi^\circ\equiv 0$. Following the terminology of [GePS], we call π hypercuspidal if φ in π implies $\varphi^\circ \equiv 0$. In particular, we are really using "non-degenerate" in this paper to mean "non hypercuspidal", a possibly stronger condition than the existence of an abstract ψ-Whittaker functional (cf. Remark 2.5(iii) on pg. 150 of [GePS]. Note that it is the constant term φ° - and not φ itself - which admits a Fourier expansion in terms of the Whittaker functions W_φ. Nevertheless, as in [GePS], this is sufficient for our purposes.

(1.2.) Method B.

The relevant data here is

$$\begin{array}{ccc} G & \subset & H \\ \cup & & \cup \\ Q & \subset & P \end{array}$$

where H and G are again reductive groups (but now G is the smaller one, and it is the Eisenstein series $E(h,s)$ which will be restricted to this smaller group), P is again a parabolic subgroup of H with Levi decomposition

$$P = GL(n)\cup^P,$$

and $Q = P \cap G$ is a (not necessarily parabolic) subgroup of G satisfying the following properties:

(1.2.1) $U^Q = \cup^P \cap G$ is normal in Q and fits into the split exact sequence

$$1 \longrightarrow U^Q \longrightarrow Q \xrightarrow{\alpha} P_n \longrightarrow 1;$$

(1.2.2) $\alpha^{-1}(Z_n)$ defines a maximal unipotent subgroup $\cup^G$ of G and $\theta \circ \alpha$ defines a non-degenerate character of U^G.

Assuming this data, we introduce a global zeta integral of the form

$$I(s,\varphi,E) = \int_{G_k \backslash G_{\mathbb{A}}} \varphi(g) E_f^\tau(g,s)\, dg$$

where φ is a cusp form on $G_{\mathbb{A}}$ in the space of π, and $E(g,s)$ is the restriction to $G_{\mathbb{A}}$ of the Eisenstein series

$$E_f^\tau(h,s) = \sum_{\gamma \in P_k \backslash H_k} f^\tau(h,s).$$

Since we want to be able to collapse the "double integration" (over $G_k \backslash G_{\mathbb{A}}$ and $P_k \backslash H_{\mathbb{A}}$) into a single integration, it will be useful to add to our axioms (1.2.1) and (1.2.2) the following property:

(1.2.3) The coset space $P \backslash H/G$ consists of one "main" G-orbit ($\approx Q \backslash G$) plus a finite number of "negligible" orbits whose stabilizers in G are parabolic subgroups.

Theorem B. With data H, G, P, Q satisfying axioms (1.2.1) - (1.2.3),

$$I(s,\varphi,E) = \int_{U_{\mathbb{A}}^G \backslash G_{\mathbb{A}}} W_\varphi(g) W^\tau(g,s)\,dh.$$

Here W belongs to the standard Whittaker model of π with respect to the non degenerate character $\psi_U = \theta \circ \alpha$ of U^G, and

$$W^\tau(h,s) = \int_{Z_n(k)\backslash Z_n(\mathbb{A})} f(zh,s)\overline{\theta(z)}\,dz.$$

Proof. Let E denote a representative for the "main" G-orbit of $P \backslash H$, and let $Eh_1, \cdots, Eh_e$ denote representatives for the remaining orbits. Let us show first that axiom (1.2.3) indeed implies that the contribution of these remaining orbits to $I(s,\varphi,E)$ is zero. By definition, the contribution of the orbit containing Eh_i is

$$\int_{G_k \backslash G_{\mathbb{A}}} \varphi(g) \sum_{\gamma \in Q_i \backslash G_k} f^\tau(h_i \gamma g, s)\,dg$$

where Q_i is the parabolic subgroup of G_k stabilizing Eh_i . Collapsing this double integration gives the contribution

$$\int_{Q_i \backslash G_{\mathbb{A}}} \varphi(g) f^i(g,s)\,dg,$$

where f^i is the function on $G_{\mathbb{A}}$ defined through the formula $f^i(g,s) = f(h_i g, s)$. Now suppose Q_i has unipotent radical U^i. Then clearly f^i is left invariant by U^i, since $f(h,s)$ is invariant by U^P. Indeed, note that the stabilizer of Eh_i in $\underline{H}$ is the parabolic subgroup $h_i^{-1}Ph_i$ of H; hence $Q_i = (h_i^{-1}Ph_i) \cap G$, and $f^i(ug,s) = f(h_i ug, s) = f(u^{h_i}h_i g, s) = f(h_i g; s) = f^i(g,s)$ for all u in U^i. The contribution to $I(s,\varphi,E)$ may therefore be rewritten in the form

$$\int_{Q_i \mathsf{U}_{\mathbb{A}}^{Q_i} \backslash G_{\mathbb{A}}} f^i(g,s) \left(\int_{\mathsf{U}_k^{Q_i} \backslash \mathsf{U}_{\mathbb{A}}^{Q_i}} \varphi(ug)du \right) dg$$

and this is zero since φ is a cusp form.

It remains to analyze the contribution from the main orbit, namely $Q \setminus G$ with $Q = P \cap G$; it is

$$I(s,\varphi,E) = \int_{Q_k \backslash G_{\mathbb{A}}} \varphi(g) f^\tau(g,s)dg. \tag{1.2.4}$$

Now recall that for each fixed h_0 in $H_{\mathbb{A}}$, the function $\beta(m) = f^\tau(mh,s)$ belongs to the space of cusp forms on $GL_n(\mathbb{A})$ realizing $\tau \otimes \|\ \|^{s'+\frac{n-1}{2}}$. So by (1.0.1) it follows that

$$f^\tau(h,s) = \sum_{\gamma \in Z_{n-1} \backslash GL_{n-1}} W^\tau(\gamma h, s) \tag{1.2.5}$$

with

$$W^\tau(h,s) = \int_{Z_n \backslash Z_n(\mathbb{A})} f^\tau(zh,s)\overline{\theta(z)}dz.$$

Note that by assumption (1.2.1), $Q = P_n \mathsf{U}^Q$. Therefore we may write Q as $GL_{n-1}U^*$, where U^* is unipotent and contains U^Q, and $Z_{n-1}\mathsf{U}^*$ equals the maximal unipotent subgroup $\mathsf{U}^G = \alpha^{-1}(Z_n)$. In particular,

$$Z_{n-1} \setminus GL_{n-1} \approx U^Q \setminus Q$$

and so substitution of (1.2.5) in (1.2.4) gives

$$I(s,\varphi,E) = \int_{U^G \backslash G_{\mathbb{A}}} \varphi(g) W_f^\tau(g,s)dg.$$

Integrating first with respect to $U_k^G \setminus \mathsf{U}_{\mathbb{A}}^G$ then gives the desired result.

(1.3.) Method C.

Our data consists now of the group G, a maximal parabolic subgroup $P = GL(n)U^P$, and a character ψ_U satisfying the following axiom:

<u>**(1.3.1)**</u> If θ denotes the standard non-degenerate character of the maximal unipotent subgroup Z_n of $GL(n)$, then the formula

$$\psi_U\theta(zu) = \psi_U(u)\theta(z)$$

defines a non-degenerate character of the (maximal) unipotent subgroup $U^G = Z_nU^P$ of G.

Because no subgroup H appears in this set up, we must introduce some other object to play the role of "restriction to the subgroup H". This role is played ultimately by a "small" representation of G, whose prototype is the Weil representation. As D. Kazhdan has observed, it should be possible to view such data as being dual to the data used in Method A; in other words, our set-up for Method C is not completely different from Method A, but rather dual to it. We shall add a few more words about this in Section 6 after we have treated the example $G = Sp(n)$ in detail. In the meantime, we simply hypothesize the existence of an automorphic representation r of $G_{\mathbb{A}}$ satisfying the following property:

<u>**(1.3.2.)**</u> Suppose $\theta(g)$ belongs to the space of r; then $\theta(g)$ enjoys a Fourier expansion of the form

$$\theta(g) = \sum_{\gamma\in P_n\backslash GL(n)} \theta_\psi(\gamma g) + F_0(g)$$

where $\theta_\psi(z_nug) = \psi_U(u)\theta_\psi(g)$ for z_n in $Z_n(\mathbb{A})$ and u in $U^P_{\mathbb{A}}$, and $F_0(g)$ is left invariant by $U^P_{\mathbb{A}}$.

Now suppose τ is an automorphic cuspidal representation of $GL_n(\mathbb{A})$. For each automorphic cuspidal representation π of $G_{\mathbb{A}}$ we consider the global zeta-integral

$$I(s,\varphi,E) = \int_{G_k\backslash G_{\mathbb{A}}} \varphi(g)\theta(g)E(g,s)dg$$

where $\varphi(g)$belongs to the space of π, and

$$E(g,s) = \sum_{P_k\backslash G_k} f^\tau(\gamma g,s)$$

is the Eisenstein series on $G_{\mathbb{A}}$ induced from $\tau \otimes |det|^{s'}$ on $P_{\mathbb{A}} = GL_n(\mathbb{A})U^P_{\mathbb{A}}$.

<u>**Theorem C.**</u> With data π, τ, G, P and r satisfying axioms (1.3.1) and (1.3.2), we have

$$I(s, \varphi, E) = \int_{U^G_{\mathbb{A}} \backslash G_{\mathbb{A}}} W_\varphi(g) \theta_\psi(g) W^\tau_f(g, s) dg.$$

As usual, we are viewing f^τ here so that for any fixed g in $G_{\mathbb{A}}$, $w(m) = f^\tau(mg, s)$ belongs to the space of cusp forms on $GL_n(\mathbb{A})$ realizing the representation $\tau \otimes |det|^{s^*}$, and $W^\tau_f(g, s) = \int_{Z_n(k) \backslash Z_n(\mathbb{A})} f(zg)\overline{\theta(z)} dz$. By $W_\varphi(g)$ we denote the Fourier coefficient of $\varphi(g)$ with respect to the non-degenerate character $\psi_U \cdot \theta$ of the (maximal) unipotent subgroup $U^G_{\mathbb{A}} = (Z_n U^P)_{\mathbb{A}}$ of $G_{\mathbb{A}}$.

<u>**Proof.**</u> Substituting in the definition of $E(g, s)$, and the Fourier expansion (1.3.1) for $\theta(g)$, gives

$$I(s, \varphi, f^\tau) = \int_{P_k \backslash G_{\mathbb{A}}} \varphi(g) F_0(g) f^\tau(g, s) dg + \int_{P_n U^P \backslash G_{\mathbb{A}}} \varphi(g) \theta_\psi(g) f^\tau(g, s) dg.$$

Since $F_0(g)$ is assumed to be left $U^P_{\mathbb{A}}$ invariant it follows immediately (by integrating first with respect to $U^P_k \backslash U^P_{\mathbb{A}}$ and recalling that φ is cuspidal) that the first integral above equals

$$\int_{P_k U^P_{\mathbb{A}} \backslash G_{\mathbb{A}}} \left(\int_{U^P_k \backslash U^P_{\mathbb{A}}} \varphi(ug) dg \right) F_0(g) f^\tau(g, s) dg = 0.$$

Substituting in the Fourier expansion for $w(m) = f^\tau(mg, s)$ at $m = 1$ then gives

$$I(s, \varphi, f^\tau_) = \int_{Z_n U^P \backslash G_{\mathbb{A}}} \varphi(g) \theta_\psi(g) W^\tau_f(g, s) dg.$$

Observe that $Z_n U^P = UG$ is a <u>maximal</u> unipotent subgroup of G. Indeed Z_n is the maximal unipotent subgroup of $GL(n)$, and $GL(n)U^P$ is a maximal parabolic subgroup of G. Integrating now first with respect to $U^G_k \backslash U^G_{\mathbb{A}}$ then gives the desired result, since $\Theta_\psi(z_n ug) = \psi_{U^P}(u) \Theta_\psi(g)$, $W^\tau_f(z_n ug) = \theta(z_n) W^\tau_f(g)$, and $\psi^U \theta$ is the correct non-degenerate character of $U^G = Z_n U^P$ used to define $W_\varphi(g)$.

§.2. Groups of type B_n : Structure Theory

Let k denote any field of characteristic not 2, V a vector space of dimension $2n+1$ over k, and $(\ ,\ ): V \times V \to k$ a non-degenerate symmetric bilinear form on V of index n. Then there exists an isotropic subspace X of V of (maximal possible) dimension n. By Witt's embedding theorem (cf. Theorem 6.11 of [Jacobson]) we can write

$$V = X \oplus (\ell_0) \oplus X^v,$$

where X^v is an isotropic subspace naturally dual to X, and ℓ_0 is an anisotropic vector orthogonal to $X \oplus X^v$. In more detail, we pick bases $\{x_1, \cdots, x_n\}$ and $\{x_1^v, \cdots, x_n^v\}$ for X and X^v so that the matrix of $(\ ,\ )$ with respect to the basis $\{x_1, \cdots, x_n, \ell_0, x_1^v, \cdots, x_n^v\}$ of V has the form

$$J = \begin{pmatrix} 0 & 0 & I_n \\ 0 & 1 & 0 \\ I_n & 0 & 0 \end{pmatrix}.$$

Now let G denote the special orthogonal group $SO(V)$, i.e.,

$$\begin{aligned} G &= \{g \in SL(V) : {}^t gJg = J\} \\ &= \{g \in SL(V) : (gx, gy) = (x,y) \forall x, y \in V\}. \end{aligned}$$

We assume, henceforth, that $n \geq 2$. Our purpose in this Section is to verify that axioms (1.1.1) and (1.1.2) are satisfied for appropriately chosen H, P, and Q.

The parabolic subgroups of G are the stabilizers of isotropic flags, i.e., up to conjugacy - flags inside X. In case the flag is a full one, we get a Borel subgroup, whose maximal torus is the k-split torus

$$\left\{ \begin{bmatrix} a_1 & & & & & & 0 \\ & \ddots & & & & & \\ & & a_n & & & & \\ & & & 1 & & & \\ & & & & a_1^{-1} & & \\ & & & & & \ddots & \\ 0 & & & & & & a_n^{-1} \end{bmatrix} : a_i \in k^x \right\}.$$

On the other hand, if the flag is just X, then we get the maximal parabolic subgroup

$$Q = \{g \in G : g(X) \subset X\}.$$

Proposition 2.1. Let $Q = MU^Q$ denote the Levi decomposition of the parabolic subgroup Q. Then

$$U^Q = \{q \in Q : q|_X = I, q|_{(X+(\ell_0))/X} = 1\}$$

and $M \approx GL_n$ via the embedding

$$g \to \begin{pmatrix} g & 0 & 0 \\ 0 & 1 & 0 \\ 0 & 0 & t_g - 1 \end{pmatrix}.$$

Proof. Implicit in our description of the unipotent radical U^Q is the fact that Q stabilizes $X + (\ell_0)$ as well as X. To verify this, suppose that q in Q is arbitrary, and $q\ell_0 = x + a\ell_0 + x^v$, with $x \in X$ $a \in k$, and $x^v \in X^v$. To show that $x^v = 0$, we note that for all $x' \in X$,

$$(\ell_0, x') = 0 = (q\ell_0, qx').$$

Thus, for any $x' \in X$, we have

$$0 = (q\ell_0, x') = (x + a\ell_0 + x^v, x') = (x^v, x'),$$

i.e., $x^v = 0$, as claimed. Note then that the condition $q|_{X+\ell_0} \equiv I$ (modulo X) is automatically satisfied for any q in Q. Indeed, $det(q) = 1$ implies that the matrix form of q is

$$\begin{bmatrix} g & * & * \\ 0 & 1 & * \\ 0 & 0 & t_g - 1 \end{bmatrix},$$

i.e, $a = 1$.

To prove that U^Q is the maximal normal unipotent subgroup of Q, consider the homomorphism $t : Q \to GL(X)$ defined by $t(q) = q|_X$. By definition, $ker\ t \supset U^Q$. On the other hand, as we have just observed that the condition $q|_{X+\ell_0} \equiv I(mod\ X)$ is redundant, we also have $ker\ t \subset U^Q$. In particular, $U^Q = ker\ t$ is normal in Q. Now we claim that the matrix form of U^Q (with respect to our previously chosen basis for V) must be

$$\left\{ \begin{pmatrix} I_n & * & * \\ 0 & 1 & * \\ 0 & 0 & I_n \end{pmatrix} \right\}.$$

Equivalently, we claim that each u in U^Q acts as the identity on X^v, at least modulo $X + (\ell_0)$. Indeed, suppose x_j^v is any one of our basis vectors for X^v. Then we compute

$$(x_i, ux_j^v) = (u^{-1}x_i, x_j^v) = (x_i, x_j^v) = \delta_{ij}$$

for any basis vector x_i in X, i.e., $ux_j^v = \sum a_{ij} x_i^v + x' + a\ell_0$, with $x' \in X$, and $a_{ij} = \delta_{ij}$. This proves that u takes each x_j^v to itself modulo $X + (\ell_0)$, as was to be shown; this matrix form also clearly establishes the unipotentcy of U^Q.

Finally, we note that our homomorphism $t : Q \to GL(X)$ is onto, since for any g in $GL(X)$, the transformation

$$\begin{pmatrix} g & 0 & 0 \\ 0 & 1 & 0 \\ 0 & 0 & t_g - 1 \end{pmatrix}$$

belongs to $Q \subset SO(V)$. Therefore Q/U^Q is isomorphic to $GL(X) = GL_n(k)$, and the maximality of U^Q among all the normal unipotent connected subgroups of Q follows from the reductivity of GL_n. □

Before bringing into play our particular Eisenstein series, and corresponding basic identity, we need to introduce a subgroup of G isomorphic to the split (special) orthogonal group in $2n$ variables. Thus we let H denote the subgroup of G which stabilizes $V' = X \oplus X^v$ and acts trivially on (ℓ_0), i.e.,

$$H = \left\{ g \in G : g = \begin{pmatrix} A & 0 & B \\ 0 & 1 & 0 \\ C & 0 & D \end{pmatrix} \right\}.$$

This subgroup is naturally isomorphic to $SO(V') = SO(2n)$ via the embedding

$$\begin{pmatrix} A & B \\ C & D \end{pmatrix} \longrightarrow \begin{pmatrix} A & 0 & B \\ 0 & 1 & 0 \\ C & 0 & D \end{pmatrix}.$$

Also $Q \cap H = P$ is a maximal parabolic subgroup of H (the one preserving the n-dimensional maximal isotropic subgroup X of V').

Henceforth, we shall confuse H and its subgroups with $SO(2n)$ and its subgroups. The condition that h acts trivially on ℓ_0 actually implies that h stabilizes V', since h is orthogonal, and $V' = (\ell_0)^\perp$.

Our purpose now is to verify axiom (1.1.1) for the data H, G, P, Q.

Proposition 2.2. Let

$$U^P = \{u \in U^Q : u|_{X+(\ell_0)} = I, \quad i.e. \quad u(\ell_0) = \ell_0\}$$
$$= U^Q \cap H.$$

Then U^P is the unipotent radical of P. Moreover, it is normal in Q, and

$$U^P \setminus U^Q \approx Hom\,(X^v, k) \approx k^n.$$

In particular, there is an exact sequence

$$1 \longrightarrow U^P \longrightarrow Q \xrightarrow{\alpha} P_{n+1} \longrightarrow 1$$

where

$$P_{n+1} = \{p = \begin{pmatrix} g & * \\ 0 & 1 \end{pmatrix} \in GL_{n+1} : g \in GL_n\}$$
$$\approx GL_n(k) \;\; k^n.$$

Proof. Recall first that

$$U^Q = \left\{ u = \begin{pmatrix} I_n & A & C \\ 0 & 1 & B \\ 0 & 0 & I_n \end{pmatrix} \right\}$$

where C is an $n \times n$ matrix, and A (respectively B) is a n-tuple column (respectively row) vector. From the fact that u belongs to $SO(2n+1)$, it is easy to check that A and B are arbitrary modulo the condition that $A +^t B = 0$; in particular, A can be taken to be arbitrary in k^n.

The condition $u(\ell_0) = \ell_0$ (which defines U^P) allows us to delete the $n+1$-th row and column from the reasoning used in Proposition 2.1; in other words, working with the natural surjection $t^P : P \to GL(X)$, we conclude now that P has Levi decomposition $GL_n(k)U^P$. Note that

$$P = Q \cap H \approx \left\{ \begin{pmatrix} A & B \\ 0 & C \end{pmatrix} \in SO(2n) \right\}.$$

Now define the map $\alpha : Q \to P_{n+1}$ by

$$\alpha \begin{pmatrix} A & * & * \\ 0 & 1 & * \\ 0 & 0 & t_A - 1 \end{pmatrix} = \begin{pmatrix} A & * \\ 0 & 1 \end{pmatrix}.$$

Since we have already observed that A in $GL(n)$ and $*$ in k^n may be chosen arbitrarily, it follows that α is onto with kernel U^P. It remains only to prove that $U^P \backslash U^Q$ is isomorphic to $Hom(X^v, k)$.

Suppose u in $U^Q$is arbitrary. Since u acts as the identity on X, the parameters of u are determined by its action on ℓ_0 and an arbitrary x^v in X^v. In particular, suppose

$$u(\ell_0) = \ell_0 + x_u, \; x_u \;\; in \;\; X$$

and

$$u(x^v) = x^v + \ell(x^v)\ell_0 + x_u' \;\; with \;\; x_u' \;\; in \;\; X.$$

Perforce, $\ell : X^v \to k$ belongs to $Hom(X^v, k)$. Because u is orthogonal, $(x^v, \ell_0) = 0 = (ux^v, u\ell_0)$ implies

$$\begin{aligned} 0 &= (x^v + \ell(x^v)\ell_0 + x_u', \ell_0 + x_u) \\ &= \ell(x^v)(\ell_0, \ell_0) + (x^v, x_u), \end{aligned}$$

i.e., $\ell(x^v) = (-x^v, x_u)$. This means that the parameters x_u and x_u' completely determine u. So consider the homomorphism

$$U^Q \longrightarrow Hom(X^v, k)$$

defined by sending u to ℓ_u, where $\ell_u(x^v) = -(x^v, x_u)$. This homomorphism is obviously onto, and its kernel consists of those u in V_Q such that $x_u = 0$, i.e., those u of such that $u(\ell_0) = \ell_0$. By definition, then, this kernel is U^P, and the proof is complete.

Remark 2.3. The radical U^P consists of matrices $\left(\begin{smallmatrix} I_n & S \\ 0 & I_n \end{smallmatrix}\right)$, where S is an arbitrary skew-symmetric matrix in $M_n(k)$. In particular U^P is abelian, and hence U^Q is two-step nilpotent.

In order to verify axiom (1.1.2), we need in particular to check that $\psi \circ \alpha$ defines a non-degenerate character of $\alpha^{-1}(Z_{n+1})$. For this, it is convenient to work with a different, yet isomorphic, realization of the special orthogonal group G, namely that belonging to the bilinear form whose matrix is

$$J' = \begin{pmatrix} 0 & & 1 \\ & \cdot^{\cdot^{\cdot}} & \\ 1 & & 0 \end{pmatrix} \quad \textit{in place of} \quad \begin{pmatrix} 0 & 0 & I_n \\ 0 & 1 & 0 \\ I_n & 1 & 0 \end{pmatrix} = J.$$

In this realization of G, the standard Borel subgroup B is the upper triangular subgroup of matrices containing the maximal torus

$$T = \left\{ \begin{pmatrix} a_1 & & & & & & 0 \\ & \ddots & & & & & \\ & & a_n & & & & \\ & & & 1 & & & \\ & & & & a_n^{-1} & & \\ & & & & & \ddots & \\ & & & & & & a_1^{-1} \end{pmatrix} : a_i \in k^x \right\}.$$

The unipotent radical of B is the maximal unipotent subgroup of matrices

$$U^G = \left\{ u = \begin{pmatrix} 1_1 & & x_{ij} \\ & \ddots & \\ & & 1 \end{pmatrix} \right\} \subset GL_{2n+1}(k)$$

where the x_{ij}'s with $1 \le i < j \le n$ are arbitrary, and the remaining entries are constrained by the condition that u belongs to $SO(2n+1)$. In particular, the x_{ij} with $n+1 \le i < j \le 2n+1$ are uniquely determined by the x_{ij} with $1 \le i < j \le n$; cf. [Borel 2.], p.16. With respect to the symmetric matrix J', the radical U^P now consists of matrices $\left(\begin{smallmatrix} I_n & S \\ 0 & I_n \end{smallmatrix}\right)$ with S skew symmetric with respect to the non-principal diagonal of $M_n(k)$.

Now fix a character ψ of k which is non-trivial, and define a character of U^G through the equation

$$\psi_U(u) = \psi\Big(\sum_{i=1}^{n} x_{i,i+1}\Big).$$

This character is non-degenerate in the sense that it is non-trivial on each of the root subgroups of U^G belonging to the n simple roots $\lambda_1 - \lambda_2, \lambda_2 - \lambda_3, \cdots, \lambda_{n-1} - \lambda_n$, and λ_n. (As usual, λ_j denotes the character of T such that $\lambda_j\ (a_1, \cdots, a_n,\ 1, a_n^{-1}, \cdots, a_1^{-1}) = a_j$. Recall that the n simple roots of $H \subset G$ are $\lambda_1 - \lambda_2, \cdots, \lambda_{n-1} - \lambda_n$ that $\lambda_{n-1} + \lambda_n$. For H or G, the root subgroups attached to the first $n-1$ simple roots are always contained in $M \approx GL_n$. However, the root subgroup of U^H belonging to $\lambda_{n+1} + \lambda_n$ lies in U^P, whereas the root subgroup of U^G belonging to λ_n lies "between" M and U^P; cf. [Borel 2], p.16.

In any case, since $\alpha^{-1}(Z_{n+1}) = \mathsf{U}^G$ and $\theta \circ \alpha$ equals ψ_U, Axiom (1.1.2) is satisfied. Thus we have:

Theorem 2.B. Suppose π(*resp.* τ) is a non-degenerate automorphic cuspidal representation of $G_{\mathbb{A}}$ (resp. $GL_n(\mathbb{A})$), φ belongs to the space of π, and $E(h,s)$ is the Eisenstein series on $H_{\mathbb{A}}$ "induced from τ". Then

$$I(s,\varphi,E) = \int_{H_k \backslash H_{\mathbb{A}}} \varphi(h) E(h,s)\,dh = \int_{U^H_{\mathbb{A}} \backslash H_{\mathbb{A}}} W_\varphi(h) W_f^\tau(h,s)\,dh$$

with $\mathsf{U}^H_{\mathbb{A}} = Z_n(\mathbb{A}) \mathsf{U}^P_{\mathbb{A}}$ the (standard) maximal unipotent subgroup of $H_{\mathbb{A}}$,

$$W_\varphi(g) = \int_{\mathsf{U}^G_k \backslash \mathsf{U}^G_{\mathbb{A}}} \varphi(ug) \psi_U^{-1}(u)\,du$$

and

$$W_f^\tau(h,s) = \int_{Z_n(k) \backslash Z_n(\mathbb{A})} f^\tau(zh,s)\theta(z)\,dz.$$

Caution. Although ψ_U is a non-degenerate character of U^G, its restriction to U^H is not. Indeed ψ_{U} is trivial on U^P, and U^P contains the root subgroup of H belonging to the simple root $\lambda_{n-1} + \lambda_n$.

§3. Groups of Type D_n : Orbits of Maximal Isotropic Subspaces

In Section 2 we established the identity

$$\int_{H_{\mathbb{A}}\backslash H_{\mathbb{A}}} \varphi(h)E_f^{\tau}(h,s)dh = \int_{U_{\mathbb{A}}^H\backslash H_{\mathbb{A}}} W_\varphi(h)W^f(h,s)dh$$

with φ a cusp form on $G = SO(2n+1)$ restricted to $H = SO(2n)$, and $E_f^{\tau}(h,s)$ an Eisenstein series on H. In this Section, we shall establish a similar identity, except that φ will now be a cusp form defined directly on $SO(2n)$, and the Eisenstein series will be one on $SO(2n+1)$ restricted to $SO(2n)$. In other words, we shall switch the roles of G and H and follow Method B. In this way, we shall (eventually) attach an L-function to representations of $SO(2n) \times GL(n)$. Originally, our plan was to treat L-functions on $SO(2n) \times GL(n)$ via the theory of the oscillator or Weil representation, as we still need to do for the symplectic groups; cf. Method C and§4. The idea of instead proceeding in the simpler way just outlined, i.e. via Method B, was directly inspired by P. Garrett's recent analysis of L-functions attached to triples of classical cusp forms: cf. his preprint [Garrett]. That Garrett's idea should work in a more general - and less classical - context was developed by S. Rallis and Piatetski-Shapiro; their ideas have important applications to the theory of L-functions for $GL(2) \times GL(2) \times GL(2)$ and related groups (see [PSR3]).

Returning to the case at hand, we need to consider Eisenstein series on $H = SO(2n+1)$ of the form

$$E(h,s) = \sum_{P_k\backslash H_k} f(\gamma h, s)$$

where f belongs to the induced space $Ind_{P_{\mathbb{A}}}^{G_{\mathbb{A}}}\tau|det|^{s'}$, and P is the maximal parabolic subgroup of H preserving some maximal isotropic subspace E of the orthogonal space V (cf. §(3.2) below). To establish our basic identity in this context, we follow Method B in analyzing G-orbits of the space $P \setminus H$. In many ways, this analysis more closely resembles the proof of the basic identity of [PSR1] than that of Section 2 of the present paper. Our goal, of course, is to verify Axioms(1.2.1) - (1.2.3).

(3.1.) As in Section 2, let V denote an orthogonal vector space of dimension $2n+1$

over the field k, and X a maximal isotropic subspace of V such that $V = X \oplus (\ell_0) \oplus X^v$. Recall that $dim\ X = n$, and $(\ell_0, \ell_0) = 1$. Let $H = SO(V)$.

Lemma (3.1.1): There is exactly one H-orbit of isotropic subspaces of V of (maximal possible) dimension n.

Proof. Given any isotropic subspace X' of V of dimension n, we want to show that there is an element h in H such that $hX' = X$. By Witt's theorem (cf. [Jacobson], p.349), we are assured of the existence of at least an orthogonal h' (in $O(V)$) taking X' to X. If $det(h') = -1$, we must still show there is a h in $SO(V)$ taking X' to X. So define h^* in $O(V)$ by $h^*(\ell_0) = -\ell_0$ and $h^* = I$ on $X \oplus X^v$. Clearly h^* preserves X and has determinant -1. Therefore $h = h^* \cdot h'$ has determinant 1, yet takes X' to X, as required.

Let I denote this single H-orbit of maximal isotropic subspaces of V. Since the form of our global zeta-integral will force us to analyze the orbits of I with respect to the proper subgroup $SO(2n) \approx G \subset H$, we fix

$$V' = X \oplus X^v \subset V,$$

and $G = SO(V')$.

Proposition 3.1.2. There are three orbits of I with respect to the action of G. Two are contained in

$$I_0 = \{E \subset I : E \subset V'\},$$

and the other is

$$I_1 = \{E \subset I : E \not\subset V'\}.$$

Proof. Clearly $I = I_0 \cup I_1$ and these two subsets are disjoint. What must be shown is that I_0 comprises two G-orbits and I_1 just one. (The orbits in I_0 will turn out to be "negligible", hence the subscript zero.)

Let $G^{\pm}$ denote the subgroup of H fixing ℓ_0 up to sign. Then $G^{\pm} \approx O(V')$. Indeed G preserves V' (the orthocomplement to (ℓ_0)) and the bilinear form restricted to V'; moreover, the natural homomorphism $G^{\pm} \to O(V')$ is onto with trivial kernel. By Witt's extension theorem, this time applied to the orthogonal space V' and orthogonal group $O(V') \cong G^{\pm}$, each X_1 in I_0 is mapped isometrically to X via some element of $G^{\pm}$. Thus I_0 comprises a single $G^{\pm}$-orbit. But G has index two in $G^{\pm}$. Thus I_0 consists of two G-orbits, say with representatives X and X°.

In order to analyze I_1, let us fix two maximal isotropic subspaces E_1 and E_2 in I_1, and try to show that there exists an g in G such that $gE_1 = E_2$. Note that any E in I_1 is such that $E' = E \cap V'$ has dimension $n-1$. (Indeed $E/E \cap V' \approx E + V'/V'$, and the assumption that $E \not\subset V'$ implies that $E + V' = V$; thus $E/E' \approx V/V'$, and hence E' is of codimension 1 in E, as claimed.)

Before continuing, we give an Example.

Example 3.1.3. Suppose $n = 2$ and E is the subspace of $V = X \oplus \ell_0 \oplus X^{v}$ spanned by x_2 and $e = x_1 + \ell_0 - x_1^{v}/2$. Then E is two-dimensional (maximal) isotropic subspace of V which is not contained in $V' = X \oplus X^{v}$, i.e. E is in I_1. In this case, $E' = E \cap V' = \{x_2\}$, and $E = \{e\} + E'$.

Returning to E_1, pick e_1 in $E_1 \setminus E_1'$. The observation of the last paragraph implies that e_1 is an isotropic vector such that $E_1 = \{e_1\} + E_1'$, and $V = \{e_1\} + V'$. Similarly, $E_2 = \{e_2\} + E_2'$. Since E_1' and E_2' are $(n-1)$-dimensional isotropic subspaces of V', Witt's theorem again gives us some g in G such that $gE_1' = E_2'$. (If $det(g) = -1$, pick an anisotropic vector e_0 in V' such that $e_0^{\perp} \supset E_2'$; then compose g with $\alpha : V' \to V'$ defined by $\alpha(e_0) = -e_0$ and $\alpha = I$ on $e_0^{\perp}$. The assumption that $n \geq 2$ is of course crucial here.)

Note that the image of e_1 under this g is again an e_2 that lies outside E_2'. The argument of the last paragraph therefore implies that we may assume $E_i = \{e_i\} + E'$, with E' the fixed space $E_1' = E_1 \cap V'$. It remains to prove then that there is a g in G which maps e_1 to e_2 and preserves E'. To find such a g, note that (e_i, ℓ_0) cannot be zero for $i = 1$ or 2. Indeed $(e_i, \ell_0) = 0$ would imply $e_i \in \ell_0^{\perp} = V'$, an obvious contradiction. Therefore we may normalize e_1 and e_2 so that $(e_i, \ell_0) = 1$. We claim now that, for $i = 1$ or 2, $e_i = \ell_0 + e_i'$

with e_i' in V' (though not in E) and $(e_i', e_i') = -1$. To see this, recall $e_i \in V = (\ell_0) \oplus V'$, and write $e_i = \alpha_i \ell_0 + e_i'$, with α_i in k and e_i' in V'. Then $1 = (e_i, \ell_0) = \alpha_i$, since $(e_i', \ell_0) = 0$, and $(e_i', e_i') = (e_i - \ell_0, e_i - \ell_0) = -2(e_i, \ell_0) + (\ell_0, \ell_0) = -1$, as claimed. Since $e_1'$and e_2' are isometric and anisotropic, we may (again by Witt) find a g in G taking e_1' to e_2'. Note, moreover, that both e_1' and e_2' are orthogonal to E'. Indeed for e' in E', $(e_i', e') = (e_i - \ell_0, e') = (e_i, e') - (\ell_0, e')$, with $(e_i, e') = 0$ since E is isotropic, and $(\ell_0, e') = 0$ since $\ell_0 \perp E'$. Thus we may find a g in G which not only takes e_1' to e_2' (and hence e_1 to e_2), but also preserves E'. This completes the proof of Proposition 3.1.2.

(3.2.) Before verifying axioms (1.2.1)-(1.2.3) we need to analyze some parabolic (type) subgroups attached to (orbits of) the various isotropic subspaces just introduced. First some notation: if G is any group, V any linear space on which G acts, and $Y \subset V$ any subspace, let S_Y^G denote the stabilizer of Y in G.

Now let E denote a representative (fixed once and for all) for the "main" G-orbit I_1 consisting of maximal isotropic subspaces of V not contained in $X \oplus X^\upsilon = V' \subset V$, and let $E' = E \cap V'$. As representatives of the remaining (negligible) orbits, we take Eh_1 and Eh_2, with $Eh_1 = X$.

By P we denote the maximal parabolic subgroup of H preserving E, i.e., $P = S_E^H$. Let $Q = S_E^G = P \cap G$.

Since E is not contained in V', Q is not a parabolic subgroup of $G = SO(V')$. On the other hand, since $E' = E \cap V'$ is an $(n-1)$-dimensional isotropic subspace of V', the subgroup

$$P^* = P(E') = S_{E'}^G$$

is indeed a parabolic subgroup of G. Note that the element g used in the proof of Prop. 3.1.2 to take e_1 to e_2 was an element of P^*. Note also that $S_X^G = S_X^H \cap G$ is a parabolic subgroup of $G = SO(V')$, namely the maximal parabolic subgroup preserving $X = Eh_1 \subset V'$. Similarly $S_{Eh_2}^G$ is parabolic, and therefore Axiom (1.2.3) is automatically satisfied in the present context.

Our task now is to verify Axioms (1.2.1) and (1.2.2). In Proposition 2.1 and 2.2 of Section 2 we described the Levi decomposition of certain parabolic subgroups Q and P.

The purpose of Proposition (3.2.2) below is to deal similarly with our new subgroups P^* and Q.

<u>**Lemma 3.2.1.**</u> As above, let E denote our fixed maximal isotropic subspace of V not contained in V' $(= X \oplus X^{\upsilon})$. Then there exists a vector e' in V' such that $(e', e') = -1$ and $e' = e - \ell_0$ for some e in $E \setminus E'$.

<u>**Proof.**</u> Let e denote <u>any</u> vector in $E \setminus E'$ so that $E = \{e\} + E'$. If necessary, we normalize e so that $e' = e - \ell_0$ belongs to V'. Then as in the proof of Proposition 3.1.2, $(e', e') = -1$.

<u>**Proposition (3.2.2).**</u> Let E, E', P^*, Q, and e' be as in the Lemma and paragraph above. Let $\mathcal{L}$ denote a two-dimensional (hyperbolic) subspace of V' which is orthogonal to $E' \oplus (E')^{\upsilon}$ and such that e' belongs to $\mathcal{L}$, i.e., $V' = E' \oplus \mathcal{L} \oplus (E')^{\upsilon}$. (In the two-dimensional example described in Example 3.1.3 we have $E' = \{x_2\}, (E')^{\upsilon} = \{x_2^{\upsilon}\}$, and $\mathcal{L} = \{x_1, x_1^{\upsilon}\}$; note that $e' = x_1 - x_1^{\upsilon}/2$ belongs to $\mathcal{L}$.) Then

(a) The Levi decomposition $M^* U^{P^*}$ of P^* is such that $M^* \approx GL_{n-1} \times SO(\mathcal{L})$; and

(b) $Q = \{g \in P^* : g(e') \equiv e' \textit{ modulo } E'\}$ and

$$Q \approx GL_{n-1} \, U^{P^*} \, .$$

(Even though Q is not a parabolic subgroup, we interpret (3.2.3) as its Levi decomposition.)

<u>**Proof.**</u> We begin by proving the first assertion in (b). Suppose $p \in Q$. Since $Q \subset H$, and every element of G acts trivially on ℓ_0, we automatically have $p\ell_0 = \ell_0$. Also $p(E) = E$, by definition of Q. Then $E = \{e\} + E'$ implies $pe = \alpha e + e''$, with $e'' \in E'$, and it remains to show $\alpha = 1$. Indeed, if $p(e) \equiv e$ *(modulo E')*, then $p(e') = p(e - \ell_0) = p(e) - p(\ell_0) \equiv e - \ell_0$ *(mod E')* $= e'$ *(mod E')* as described. To show $\alpha = 1$, we compute the inner product

$$(p(e), \ell_0) = \alpha(e, \ell_0) + (e'', \ell_0) = \alpha(e, \ell_0)$$

since E' (and hence e'') is orthogonal to ℓ_0. On the other hand, p in G also implies

$$(p(e), \ell_0) = (e, p^{-1}(\ell_0)) = (e, \ell_0),$$

so we must have $\alpha = 1$. This proves

$$Q \subset \{g \in P^* : g(e') \equiv e' \; mod \; E'\},$$

and the reverse direction is obvious.

For part (a), we proceed as in the proof of Prop.2.1 of Section 2. First we argue analogously to show that p in P^* stabilizes $E' \oplus \mathcal{L}$ as well as E'. Then we consider the epimorphism $t^* : P^* \to GL(E') \times SO(\mathcal{L})$ defined by $t(p) = p|_{E'} \times A_p$, where $p\ell = x + A_p(\ell), \ell \in \mathcal{L}$, $x \in E'$, and $A_p(\ell) \in \mathcal{L}$. (The fact that $A_p \in SO(\mathcal{L})$ follows from the computation $(\ell, \ell) = (p\ell, p\ell) = (x + A_p\ell, x + A_p\ell) = (A_p\ell, A_p\ell)$, etc.) Just as in Prop. 2.1, the unipotent radical U^{P^*} will be the kernel of t^*. Since E' is $(n-1)$ dimensional, this shows $M^* \approx GL_{n-1} \times SO(\mathcal{L})$, as claimed in (a).

To complete the proof of (b), note that $SO(\mathcal{L})$ itself has no non-trivial unipotent elements. From the description of Q given in the first part of (b) it then follows that $Q \approx GL_{n-1}U^{P^*}$.

We are now prepared to analyze a certain non-degenerate character of the maximal unipotent subgroup of G.

Let Z_n denote "the" maximal unipotent subgroup of GL_n (uniquely determined only up to conjugation in GL_n). Each non-trivial character ψ of $k \setminus \mathbb{A}$ may be regarded as a (nondegenerate) character of Z_n by identifying Z_n with the group of matrices

$$\left\{ \begin{pmatrix} 1 & & & \\ & 1 & & z_{ij} \\ & & \ddots & \\ 0 & & & 1 \end{pmatrix} \right\}$$

and defining $\psi((z_{ij})) = \psi(\sum_{i=1}^{n-1} z_{i,i+1})$. Because GL_n embeds as a subgroup of G, so does Z_n. However, this embedding of GL_n in G depends on our choice of maximal isotropic subspace E of V, and differs from the familiar embedding

$$g \longrightarrow \begin{pmatrix} g & 0 \\ 0 & {}^t g - 1 \end{pmatrix}$$

which was associated to the isotropic subspace X (and basis $x_1, \cdots, x_n, x_1^v, \cdots, x_n^v$).

The Proposition below describes just how (and why) we choose a particular embedding of Z_n in G.

Proposition 3.3. Axioms (1.2.1) and (1.2.2) hold in the present context. More precisely, recall that $P \approx GL(n) \cup^P$ is the stabilizer of E in H, and $Q = P \cap G$. Let $U^Q = U^P \cap G \subset Q$. Then there is a split exact sequence $1 \to \cup^Q \to Q \overset{\alpha}{\to} P_n \to 1$, and a choice of embedding of Z_n in GL_n and in P such that

(a) $\alpha^{-1}(Z_n) = Z_n \cup^Q$ is a maximal unipotent subgroup U^G of G; and

(b) if $\psi : Z_n \to \mathbb{C}$ is a non-degenerate character, then the formula

$$\tilde{\psi}(z_n u') = \psi(z_n), z_n \in Z_n, u' \in U^Q,$$

defines a non-degenerate character of U^G.

Proof. Our first step will be to reinterpret the Levi-type decomposition (3.2.3) for the (non-parabolic) subgroup Q of G. For this, recall the decomposition

$$V = (\ell_0) \oplus V',$$

where $V' = X \oplus X^v$, and $\ell_0 \perp V'$. Recall that $E' = E \cap V'$ is an $n-1$ dimensional subspace of E. As in Lemma 3.2.1 we fix the vector $e = e' + \ell_0$ in $E \setminus E'$. Recall that $(e, \ell_0) = (e' + \ell_0, \ell_0) = 1$, since $(e', \ell_0) = 0$. As in the proof of Proposition 3.2.2, we conclude that $re \equiv e$ (*modulo* E') for any r in Q. Indeed the fact that E has codimension 1 in E implies that $r(e) \equiv \lambda e \bmod E'$. So since any r in R preserves ℓ_0, we have $(r(e), \ell_0) = \lambda(e, \ell_0) = (e, r^{-1}(\ell_0)) = (e, \ell_0)$, i.e., $\lambda = 1$, as claimed.

Now let

$$P_n = \{p \text{ in } GL(E) : pe \equiv e(E')\},$$

and consider the natural homomorphism

$$\alpha : Q \longrightarrow GL(E)$$

given by $\alpha(r) = r|_E$. According to our observation above, $\alpha(Q) \subset P_n$. What we claim now is that $\alpha(Q) = P_n$. In fact, we shall show that there is a subgroup of Q which is isomorphic via α to P_n (and which we again denote by P_n). Since the kernel of α is clearly U^Q, the conclusion will be that $Q = P_n U^Q$.

Given p in $P_n \subset GL(E)$, our task is to extend p to all of V in such a way that this extension belongs to $Q \subset G$. Recall first that $e = \ell_0 + e'$ with e' in V' such that $(e', e') = -1$. Recall also the orthogonal decomposition $V' = E' \oplus \mathcal{L} \oplus (E')^v$, where $\mathcal{L}$ is a two-dimensional hyperbolic plane in V' containing e'. Write $\mathcal{L}$ as the orthogonal sum of e' and e''. Our strategy is to extend p in $P_n \subset GL(E)$ to all of V by defining it properly on e', viz. on E we define this extension r_p to agree with p, and we set

$$r_p(e') = e' + p(e) - e.$$

It is easy to check that r_p must then preserve ℓ_0. Indeed $r(\ell_0) = r(e - e') = p(e) - r(e') = p(e) - (e' + p(e) - e) = e - e' = \ell_0$. On $(E')^v$ we define r_p in the natural way, and we set $r_p(e'')$ equal to e''. Thus r_p extends to an element of $SO(V)$ belonging to G (since it preserves ℓ_0); moreover, as p runs through P_n, the resulting set of transformations r_p comprises a subgroup of Q isomorphic to P_n via α.

Now, finally, we can describe the required embedding of Z_n in P. Fix any basis $\{e_1, \cdots, e_{n-1}\}$ for $E' \subset E$, and let $E_i = \{e_1, e_2, \cdots, e_i\}, i = 1, \cdots, n-1$. Then let Z_n be the unipotent stabilizer of the full flag

$$E_1 \subset E_2 \subset \cdots \subset E_{n-1} \subset E,$$

i.e., $Z_n = \{z \in GL(E) : z|E_i = I \text{ } modulo \text{ } E_{i-1}\}$. Note $Z_n \subset P_n$ since $ze \equiv e \text{ } modulo \text{ } E_{n-1} (= E')$. Let Z_n also denote the subgroup of Q isomorphic to $Z_n \subset P_n$ via the above splitting of $\alpha : Q \to P_n$. Then $Z_n U^Q \subset Q$. That $Z_n U^Q$ is a maximal unipotent subgroup of G follows from the fact that Z_n is a maximal unipotent subgroup of P_n. This completes the proof of Proposition 3.3 through Part (a).

<u>**Remark 3.3.1.**</u> Because $P^* = GL_{n-1} SO(\mathcal{L}) U^{P^*} \supset Q = GL_{n-1} U^{P^*} = P_n U^Q$, and because $Z_{n-1} \subset GL_{n-1}$ is a maximal unipotent subgroup, it follows that $Z_{n-1} U^{P^*}$ is also a maximal unipotent subgroup of G. Indeed $Z_n U^Q = Z_{n-1} U^{P^*}$, a fact we shall soon

exploit in the proof of our basic identity.. Henceforth, we denote this maximal unipotent subgroup of G by U^G.

To prove part (b) of Prop. 3.3, we need to analyze the maximal unipotent subgroups of G in a general context. Suppose then that W is any $2n$-dimensional vector space with symmetric bilinear form $(\ ,\ )$, for example, $W = V' \subset V$. Let L denote any maximal isotropic subspace of dimension n (e.g. $L = X$) and set $H = SO(W)$. Write $W = L \otimes L^V$, with L^v the dual of L. Take a full flag $\{0\} = L_0 \subset L_1 \subset L_2 \subset \cdots \subset L_n = L$, and consider the sequence of nested subspaces

$$L \subset L_{n-1}^{\perp} \subset L_{n-2}^{\perp} \subset \cdots \subset L_1^{\perp} \subset W = L_0^{\perp}.$$

The unipotent stabilizer of this flag is a typical maximal unipotent subgroup Z of H, and the corresponding simple root subgroups $Z^i, i = 1, \cdots, n$, may be described as follows.

For each $i \leq n-1$, let α_i denote the character $\alpha(t_1, \cdots, t_n, 1, t_n^{-1}, \cdots, t_1^{-1}) = t_i/t_{i+1}$, and let α_n denote the character $\alpha(t_1, \cdots, t_1^{-1}) = t_{n-1}t_n$. Then Z^i is a one-dimensional subgroup of Z, together with an isomorphism $x_i : k \to Z^i$ (uniquely determined up to multiplication by a non-zero scalar) such that $tx_i(\lambda)t^{-1} = x_i(\alpha_i(t)\lambda)$ for all t in the maximal torus T; note that x_i^{-1} defines a homomorphism $Z^i \to k$. Alternatively, each Z^i may also be described in terms of the flag L_i defining Z. For each $i \leq n-1$, let ρ_i denote the natural homomorphism from Z to k given by the action of Z in the two-dimensional space $L_{i-1} \setminus L_{i+1}$. Then for these i, Z^i is the subgroup of Z such that $\rho_i|_{Z^i}$ is an isomorphism onto k, and Z^n is the subgroup of Z which stabilizes $L_{n-1}^{\perp}$ and acts as the identity on L_n.

An arbitrary homomorphism $\rho : Z \to F$ is called non-degenerate if $\rho^i = \rho|_{Z^i}$ is an isomorphism for all $i \leq n$. A character $\psi_Z : Z \to \mathbb{C}$ is non- degenerate if it equals $\psi^\circ\rho$ for some nondegenerate $\rho : Z \to F$ and some non-trivial ψ.

Example 3.3.2. Suppose $n = 2$. Then W is four dimensional, say $W = \{x_2, x_1, x_{-1}, x_{-2}\}$, with $(x_{-i}, x_i) = 1$ and $(x_k, x_\ell) = 0$ if $k+\ell \neq 0$. Let $L_1 = \{x_2\}$ and $L_2 = \{x_1, x_2\}$. Then $L_1^{\perp} = \{x_2, x_1, x_{-1}\}$, $L_0^{\perp} = W$, and the maximal unipotent subgroup

Z stabilizing this flag operates on W as follows:

$$\begin{cases} x_2 \longrightarrow x_2 \\ x_1 \longrightarrow x_1 + ux_2 \\ x_{-1} \longrightarrow x_{-1} + ax_1 + wx_2 \\ x_2 \longrightarrow x_2 + bx_{-1} + cx_1 + dx_2 \end{cases}.$$

In order that z acts orthogonally on these basic elements, we must have $a = 0$, $b = -u$, $c = -w$, and $d = -uw$, i.e. each z in Z has matrix form

$$\begin{pmatrix} 1 & u & w & -uw \\ 0 & 1 & 0 & -w \\ 0 & 0 & 1 & -u \\ 0 & 0 & 0 & 1 \end{pmatrix}.$$

Moreover, it is clear that $x_1^{-1}(z) = u$ and $x_2^{-1}(z) = w$. Thus non-degenerate characters of Z are of the form $\psi_Z(z) = \psi(\alpha u + \beta w)$ with $\alpha, \beta \neq 0$.

Now suppose $\alpha = \sum_{i=1}^{n-1} a_i\alpha_i$ is the nondegenerate homomorphism of the group of upper unipotent matrices

$$\left\{ \begin{bmatrix} 1 & & z_{ij} \\ & \ddots & \\ 0 & & 1 \end{bmatrix} \right\}$$

defined through the formula $\alpha(z_{ij}) = \sum_{i=1}^{n-1} a_i z_{i,i+1}$, with $a_i \neq 0$ for all i. Then to prove part (b) of Prop. 3.3 it remains to prove that the formula $\rho(zu') = \alpha(z)$ defines a non-degenerate homomorphism of $Z_n U^Q$. Let us first suppose $n = 2$. In the example above we have computed how any element of the maximal unipotent subgroup acts on the basis elements $x_2^\vee, x_1^\vee, x_1, x_2$ (and also ℓ_0, which it preserves). The question is, what is the corresponding action of Z_2? Here Z_2 is embedded in H via the choice of flag $E' \subset E$, with $E' = \{x_2\}$, and $E = \{x_2, \ell_0 + v'\}$. Without loss of generality, we may assume $v' = ax_1 + bx_1^\vee$ with $ab = -1/2$; in particular, neither a nor b is 0. Then it is easy to compute that $x_2 \to x_2$ and

$$\ell_0 + v' \to \ell_0 + a(x_1 + ux_2) + b(x_1^\vee + wx_2)$$

$$= \ell_0 + v' + (au + bw)x_2.$$

Thus $\rho(z_2u') = \alpha\begin{pmatrix} 1 & au+bw \\ 0 & 1 \end{pmatrix} = a_1(au + bw)$, with $a_1, a, b \neq 0$. This proves the non-degeneracy of ρ, as desired.

For arbitrary n, we observe that the proof may be reduced to the case $n = 2$ just established. Indeed, for each $k = 1, \cdots, n-2$, we have $E_{k+1} \subset E' \subset L$, a maximal isotropic subspace of V', so we may choose a flag in L such that $L_{k-1} \setminus L_{k+1} = E_{k-1} \setminus E_{k+1}$. Thus

$$\rho_1 = \alpha_1, \rho_2 = \alpha_2, \cdots, \rho_{n-2} = \alpha_{n-2},$$

and matters are reduced to analyzing only ρ_{n-1} and ρ_n, more precisely, to proving that

$$(*) \qquad \alpha_{n-1} = a\rho_{n-1} + b\rho_n, \quad with \quad a, b \neq 0.$$

The derivation of (*), however, concerns only the four-dimensional orthogonal group of $L_{n-2} \setminus L_{n-2}^{\perp}$ (properly embedded in $SO(5)$), and amounts to the computation just made for the case $n = 2$. Thus the proof of Proposition 3.3 is complete.

From Theorem B we then get:

Theorem 3.B. For each $f^{\tau}(h, s)$ in the space $ind_P^H \tau \|_{\mathbb{A}}^{s'}$, set

$$(3.3.3) \qquad W^{\tau}(h, s) = \int_{Z_n(k) \setminus Z_n(\mathbb{A})} f(zh, s)\overline{\theta(z)}dz.$$

Then, for $Re(s) >> 0$, and φ a cusp form in the space of π on $SO(2n)_{\mathbb{A}}$,

$$\int_{G_k \setminus G_{\mathbb{A}}} \varphi(g) E_f^{\tau}(g, s) dg = \int_{U_{\mathbb{A}}^G \setminus G_{\mathbb{A}}} W_{\varphi}^{\psi}(h) W^{\tau}(h, s) dh.$$

Here W_{φ}^{ψ} belongs to the standard Whittaker model of π with respect to the non-degenerate character of U^H described in Proposition 3.3.

§4. Groups of Types C_n : The Weil Representation

As already suggested in our description of Method C in Section 1.3, the form of our basic integral for the symplectic group will be

(4.0.1)
$$\int_{G_k \backslash G_{\mathbb{A}}} \varphi(g)\Theta(g)E(g,s)dg$$

where $G = Sp(n)$ and $\Theta(g)$ is a theta-function in the space of the Weil representation r. Since the proper home for Θ is not really G, but rather a certain covering group of G, we begin by recalling some relevant facts from the theory of theta-functions and the metaplectic group.

Note first that the intregral (4.0.1) reduces to the integral studied in [GeJa] in case $n = 1$; indeed, in this case the L-function on $G \times GL(n)$ reduces to the 3-dimensional "adjoint square" L-function analyzed therein. For arbitrary n, we at least begin our discussion by generalizing the description of the metaplectic group given in [GeJa].

(4.1) We let V denote a vector space of dimension $2n$ over k, equipped with an antisymmetric form $(\ ,\ )$ and a maximal isotropic subspace X of V of dimension n. As usual, we write $V = X \oplus X^v$, where X^v is also isotropic, and we chose a basis $\{x_1, \cdots, x_n\}$ (resp. $\{x_1^v, \cdots, x_n^v\}$) for X (resp. X^v) so that the corresponding matrix of the form $(\ ,\ )$ is

$$J = \begin{pmatrix} 0 & I_n \\ I_n & 0 \end{pmatrix}.$$

By G we denote the symplectic group $Sp(V) = \{g \in GL(V) : {}^t gJg = J\}$. The parabolic subgroups of G are the stabilizers of flags inside X. The Borel subgroup which preserves the full flag $< x_1 > \subset < x_1, x_2 > \subset \cdots \subset V$ has the form

$$B = \left\{ \begin{bmatrix} a_1 & & * & \cdots & & * \\ & \ddots & & & & \vdots \\ & & a_n & & & \\ & & & a_1^{-1} & & \\ & & & & \ddots & \\ 0 & & & & & a_n^{-1} \end{bmatrix} \right\} = TU.$$

On the other hand, the stabilizer of X is the maximal parabolic subgroup P with Levi component

$$M = \left\{ \begin{bmatrix} g & 0 \\ 0 & t_g - 1 \end{bmatrix} \right\}$$

and (abelian) unipotent radical

$$U^P = \left\{ \left[\begin{pmatrix} I_n & X \\ 0 & I_n \end{pmatrix} \right] : {}^tX = X \right\}.$$

(4.2) Henceforth, we let k denote an $\mathbb{A}$-field not of characteristic 2, and $\mathbb{A}$ its ring of adeles. For each place v of k, except $k_v \approx \mathbb{C}$, the (local) metaplectic group $Mp(k_v)$ is a group of unitary operators on $L^2(k_v^n)$ which fits into the exact sequence

(4.2.1) $$1 \to \mathbb{T} \to Mp(k_n) \xrightarrow{pr} G_v \to 1.$$

Here $G_v = Sp(V_v)$, and $\mathbb{T} = \{z \in \mathbb{C} : z\bar{z} = 1\}$ is regarded as the group of unitary scalar operators $\Phi \to \lambda\phi$ in $L^2(k_v^n)$; it is central in Mp. The sequence (4.2.1) splits over the (local) subgroups M and $\cup^P$. If $g \cdot \Phi$ denotes the image of Φ in $L^2(k_v^n)$ under g in $Mp(k_v)$, the splitting homomorphisms $r_v : \cup^P \to Mp$ and $r_v : M \to Mp$ are defined by

(4.2.2) $$[r_v(u)] \cdot \Phi(x) = \psi(xs^tx)\Phi(x)$$

if $u = \begin{bmatrix} 1 & s \\ 0 & 1 \end{bmatrix} \in \cup_v^P$, and

(4.2.3) $$[r_v(m)] \cdot \Phi(x) = |det\ g|^{1/2}\Phi(xg)$$

if $m = \begin{bmatrix} g & 0 \\ 0 & t_g - 1 \end{bmatrix}$.

Also, we define w_0 in $Mp(k_v)$ by $w_0 \cdot \Phi(x) = \hat{\Phi}(-x)$, where $\hat{\Phi}$ is the Fourier transform of Φ with respect to the character ψ, so that $pr(w_0) = w = \begin{bmatrix} 0 & -I_n \\ I_n & 0 \end{bmatrix}$.

Globally, the metaplectic group is a group of unitary operators in $L^2(k_{\mathbb{A}}^n)$ which fits into the central exact sequence

(4.2.4) $$1 \to \mathbb{T} \to M_p(\mathbb{A}) \to G_{\mathbb{A}} \to 1,$$

where $G_{\mathbb{A}} = Sp(V_{\mathbb{A}})$. We again write $g \cdot \Phi$ for the image of Φ by the unitary operator g. According to paragraph 41 of [Weil], the sequence (4.2.4) splits over the subgroup

$G_k = Sp(V)$, as well as over $U^P_{\mathbb{A}}$ and $M_{\mathbb{A}}$; the splitting homomorphism r_k is determined by the condition

(4.2.5)
$$\sum_{\xi \in k^n} \Phi(\xi) = \sum_{\xi \in k^n} [r_k(\gamma)]\Phi(\xi), \quad \Phi \in S(\mathbb{A}^n_k),$$

where $S(\mathbb{A}^n_k)$ denote the Schwartz-Bruhat space $\Pi_v S(k^n_v)$. Also, if $w_0 \cdot \Phi(x) = \hat{\Phi}(-x)$, then $pr(w_0) = w = \begin{bmatrix} 0 & -I_n \\ I_n & \end{bmatrix}$ and $r_k(w) = w_0$.

Finally, there is a homomorphism $\xi_v : Mp(k_v) \to Mp(\mathbb{A})$ determined by the condition

$$\xi_v(g_v) \cdot \Phi(x) = (g \cdot \Phi_v)(x_v)\Pi_{w \neq v}\Phi_w(x_w)$$

for each $g_v \in Mp(F_v)$ and $\Phi = \Pi_w \Phi_w$ in $S(\mathbb{A})$, so that the diagram

$$\begin{array}{ccc} Mp(k_v) & \xrightarrow{pr_v} & G_v \\ \downarrow \xi_v & & \downarrow i_v \\ Mp(\mathbb{A}) & \longrightarrow & G_{\mathbb{A}} \end{array}$$

commutes. We recall that if k_v does not have residual characteristic 2, and the conductor of ψ_v is O_v, then (4.2.1) also splits over the maximal compact subgroup $K_v = Sp_n(O_v)$. In particular, if r_{K_v} denotes this splitting homomorphism, and Φ^0_v denotes the characteristic function of $\Pi^n_{i=1}O_v$, then $[r_{K_v}(k)] \cdot \Phi^0_v = \Phi^0_v$ for all k in K_v.

Now given $h = (h_v)$ in $G_{\mathbb{A}}$, we choose g_v in $Mp(h_v)$ so that $pr_v(g_v) = h_v$ for all v, and $g_v = r_{K_v}(h_v)$ for almost all v ; then there is exactly one g in $Mp(\mathbb{A})$ such that

(4.2.6)
$$g \cdot \Phi(x) = \Pi g_v \cdot \Phi_v(x_v)$$

and $pr(g) = h$. In this case, we write (g_v) for g even though $Mp(\mathbb{A})$ is not a restricted product of the groups $Mp(k_v)$. It is this realization of $Mp(\mathbb{A})$ as a group of operators in $S(\mathbb{A}^n_k)$ that we call "the Weil representation of $G_{\mathbb{A}}$ "; by abuse of notation, we henceforth write $r(g)\Phi$ in place of $g \cdot \Phi$, for g in $Mp(\mathbb{A})$.

(4.3) Having introduced the Weil representation, we now bring into play its raison d'étre - the basic theta-functions. For each Φ in $S(\cdot\mathbb{A}^n_k)$, and g in $Mp(\mathbb{A})$, set

(4.3.1)
$$\Theta_\Phi(g) = \sum_{\xi \in k^n} r(g)\Phi(\xi).$$

Since $r(\lambda g)\Phi = \lambda r(g)\Phi$ for $\lambda \in \mathbb{T}$, this theta-function is a "genuine" function on $Mp(\mathbb{A})$. By its very definition, it is invariant on the left by $r_k(G_k)$; it is also "slowly increasing" (cf. [Weil]).

Lemma (4.3.2) Let e_n denote the row vector $(0 \cdots 0\ 1)$ in k^n. Define $F(g)$ on $Mp(\mathbb{A})$ by $F(g) = r(g)\Phi(e_n)$, and $F_0(g)$ by $r(g)\Phi(0)$. Then

$$\Theta_\Phi(g) = F_0(g) + \sum_{\gamma \in P_n U \backslash P} F(\gamma g). \tag{4.3.3}$$

Proof. The natural linear action of $GL(n)$ on $X_n \approx k^n$ has two orbits: O and the orbit of e_n. Moreover, the isotropy group of e_n in $GL(n)$ is P_n. Therefore the Lemma follows from the definition of $\Theta_\Phi(g)$, formula (4.2.3), and the fact that $P \approx GL(n)U^P$. (Note that $|det\ g|_{\mathbb{A}} = 1$ when $det\ g$ is a principal idele.)

Henceforth, we denote by $\tilde{H}$ the inverse image in Mp of any subgroup H of G with respect to the projection $pr : Mp \to G$. In order to relate the simple Lemma (4.3.2) to our basic integral, we need to recall more of the structure theory of G.

To this end, recall that the root subgroups of G corresponding to the simple roots of G with respect to T all lie in the Levi component M of P except for one. More precisely, using the familiar notation $\{\lambda_1, \cdots, \lambda_n\}$ for the canonical basis of the character group of T, the simple roots of G are $\{\alpha_i = \lambda_i - \lambda_{i+1} : i = 1, \cdots, n-1\} \cup \{\alpha_n = 2\lambda_n\}$, and only the root group of $2\lambda_n$ lies outside M.

Now let $\rho_n : U \to k$ be the (additive) homomorphism corresponding to the root group of $2\lambda_n$, i.e.

$$\rho_n\left(\begin{bmatrix} I_n & S_{ij} \\ 0 & I_n \end{bmatrix}\right) = s_{nn}.$$

Lemma 4.3.4. Let R denote the subgroup $P_n U$ of $P \approx GL(n)U^P$, and define a character ψ_U on U^P through the formula

$$\psi_U(u) = \psi(\rho_n(u)).$$

Then $F(g) = r(g)\Phi(e_n)$ is invariant with respect to the rational points $R_k \subset P_k$, and is a ψ_U-eigenfunction with respect to $(Z_n)_{\mathbb{A}} U^P_{\mathbb{A}}$.

Proof. Suppose $p \in (P_n)_k$ and $u \in U^P_{\mathbb{A}}$. Then by formula (4.2.3),

$$F(pug) = r_\psi(p)(r_\psi(ug)\Phi)(e_n)$$
$$= r_\psi(ug)\Phi(e_n),$$

since P_n stabilizes e_n, and both μ_ψ and $||_{\mathbb{A}}$ are trivial on principal ideles. On the other hand, by formula (4.2.2), we have $(r_\psi(ug)\Phi)(e_n) = \psi(e_n S^t e_n)(r_\psi(g)\Psi)(e_n) = \psi(\rho_n(u))(r_\psi(g)\Psi)(e_n) = \psi_U(u)F(g)$, if $u = \begin{bmatrix} I & S \\ 0 & I \end{bmatrix}$. So since ψ is trivial on rational ideles, this proves $F(g)$ is left invariant by the "rational" points of R.

Now suppose $z \in (Z_n)_{\mathbb{A}}$ and $u = \begin{bmatrix} I & S \\ 0 & I \end{bmatrix} \in U^P_{\mathbb{A}}$. Because the determinant is trivial on $(Z_n)_{\mathbb{A}}$, formulas (4.2.2) and (4.2.3) imply that

$$F(zug) = \psi(\rho_n(u))(r_\psi(g)\Phi)(e_n)$$
$$= \psi_U(u)F(g),$$

and the proof of Lemma (4.3.4) is complete.

Note that the formulas above imply that $F_0(g) = r(g)\Phi(0)$ is actually left-invariant by $U^P_{\mathbb{A}}$. Therefore the last two Lemmas imply that Axiom (1.3.2) for Method C is satisfied in the present context.

Lemma 4.3.5. Let $U^G = Z_n U^P$, where Z_n is embedded in$P_n \subset GL_n \subset G$. Let θ denote the "standard" non-degenerate character

$$\theta(z) = \psi(\sum_{i=1}^{n-1} z_{i,i+1})$$

of Z_n. Then U^G is a maximal unipotent subgroup of G, and $\theta\psi_U$ is a non-degenerate character of U^G. i.e., Axiom (1.3.1) is also satisfied in the present context.

Proof. The first statement follows from the fact that Z_n is the maximal unipotent subgroup of the Levi compent M of the parabolic subgroup P of G, and U^P is the unipotent radical of P. The second statement is immediate from the description of simple roots at the beginning of (4.3).

(4.4.) Having verified Axioms(1.3.1) and (1.3.2) in the present context, it remains only to define $E(g,s)$ in the integrand of (4.0.1) so that the integral makes sense. The problem is that $\Theta_\Phi(g)$ is a genuine function on $Mp(\mathbb{A})$, whereas the integration in (4.0.1) is taken with respect to $G_\mathbb{A}$.

We note that if φ_1 and φ_2 are any two genuine functions on $Mp(\mathbb{A})$, then the product $\varphi_1\varphi_2$ is invariant by $\mathbb{T}$; thus there is a function f on $G_\mathbb{A}$ such that $\varphi_1\varphi_2(g) = f(pr(g))$ (and hence $\varphi_1\varphi_2$ may be confused with f). Our task then is to define a genuine Eisenstein series $E^\tau(g,s)$ so that the function $\Theta_\Phi(g)E^\tau(g,s)$ may be regarded as a slowly increasing function on $G_\mathbb{A}$.

Suppose τ is an automorphic (nondegenerate) cuspidal representation of $GL_n(\mathbb{A})$, and $K_\mathbb{A}$ is the standard maximal compact subgroup of $G_\mathbb{A}$. Suppose $f^\tau(g,s)$ is a continuous function of $\mathbb{C} \times Mp(\mathbb{A})$ so that (as a function on $Mp(\mathbb{A})$) f^τ is genuine, left invariant by $U^P_\mathbb{A}$, and right $\tilde{K}_\mathbb{A}$-finite; moreover, suppose that for each g in $Mp(\mathbb{A})$, the function $w(m) = f^\tau(mg,s)$ belongs to the space of cusp forms on $M_\mathbb{A}$ realizing $\tau \otimes |det\ m|^{s'+\frac{n-1}{2}}$. The Eisenstein series corresponding to f^τ is

$$E^\tau(g,s) = \sum_{\gamma\in P_k\backslash G_k} f^\tau(\gamma g,s).$$

Because $|f^\tau|$ may be regarded as a function on $G_\mathbb{A}$, the usual arguments imply that the series defining $E^\tau(g,s)$ converges for $Re(s) >> 0$.

Now consider the basic integral

$$I(s,\varphi,E) = \int_{\mathbb{T}G_k\backslash Mp(\mathbb{A})} \varphi(pr(g))\Theta_\Phi(g)E^\tau(g,s)dg$$

for $Re(s) >> 0$. For all practical purposes, we may manipulate the integrand here as if it had initially been defined as an ordinary function on $G_k \setminus G_{\mathbb{A}}$. In particular,

$$\begin{aligned}
I(s,\varphi,E) &= \int \varphi(pr(g)) \sum f^{\tau}(g,s)\Theta_{\Phi}(g)dg \\
&= \int_{\mathbb{T}P_k\setminus Mp(\mathbb{A})} \varphi(pr(g)) f^{\tau}(g,s)\Theta_{\Phi}(g)dg \\
&= \int_{P_k\setminus G_{\mathbb{A}}} \varphi(g)F_0(g)f^{\tau}(g,s)dg \\
&+ \int_{P_n U^P\setminus G_{\mathbb{A}}} \varphi(g)F(g)f^{\tau}(g,s)dg
\end{aligned}$$

Applying Theorem C therefore yields :

Theorem 4.C. For each f^{τ} as above, let

$$W^{\tau}(g,s) = \int_{Z_n\setminus Z_n(\mathbb{A})} f^{\tau}(vg,s)\bar{\theta}(v)dv$$

with θ the non-degenerate character of Z_n defined in Lemma 4.3.5. Then for $Re(s) >> 0$,

$$I(s,\varphi,E) = \int W^{\psi}(g)F(g)W_f^{\tau}(g,s)dg$$

with $F(g) = r(g)\Phi(e_n)$ as in Lemma (4.3.2), and $W^{\psi}(g)$ the Fourier coefficient of $\varphi(g)$ with respect to the character $\theta\psi_U$ of the maximal unipotent subgroup U^G of G.

Concluding Remarks: (a) The basic integral (4.0.1) is yet another generalization of a "Shimura type" zeta-integral. The original such integral first appeared - in classical form - in [Shimura]. As already mentioned in the introduction to this Section, in terms of group representations it appeared adellically in [Ge Ja] in the form

$$\int \varphi_{\pi}(g)\Theta(g)E(g,s)dg,$$

with $\varphi_{\pi}(g)$ a cusp form on $SL_2(\mathbb{A})$, and Θ and E "genuine" automorphic forms on the metaplectic covering group of $SL(2)$. A slight variant of this integral involves switching the roles of φ and E so that φ is a cusp form on the metaplectic group and E is an ordinary Eisenstein series on $SL(2)$. This integral was also first introduced by Shimura - in fact in the earlier paper [Shimura 2], and then generalized in [Ge PS3] in order to attach

L-functions to cusp forms on the metaplectic group. In the present context of $Sp(n)$, it might be of interest to develop the analogue of this integral, i.e., modify (4.0.1) so that φ is a cusp form on the metaplectic cover of $Sp(n)$ and $E(g,s)$ is an ordinary Eisenstein series on $Sp(n)$.

(b) A still more general form of Method C ("Shimura's method") is outlined in [PS 2], Part III, pp. 588 - 589. This general form of the method is directed at the case when π need not have a standard Whittaker model. In the present context, where the unipotent radical of P is abelian, conditions (1) - (3) on p.588 of [PS 2] simplify considerably.

§5. Euler Product Expansions

(5.1) We need to proceed from a basic identity of the form

$$\int_{H_k\backslash H_{\mathbb{A}}} \varphi(h)E_f^{\tau}(h,s)dh = \int_{U_{\mathbb{A}}^H\backslash H_{\mathbb{A}}} W_{\varphi}(h)W_f^{\tau}(h,s)dh$$

to an analysis of L-functions on the group $G\times GL(n)$. Thus we must factor the right-hand side of this identity as the product of local zeta integrals over the places of k. Since the adelic domain of integration obviously factors, it remains only to show that:

(1) φ and f^{τ} can be so chosen that the integrand $W_{\varphi}(h)W_f^{\tau}(h)$ also factors, say as the product of local functions $W_v(h_v)W_v^f(h_v,s)$; and

(2) the resulting local zeta-integrals

$$\varsigma(s,W_v,f_v) = \int_{U_v\backslash H_v} W_v(h)W_v^f(h,s)dh$$

converge absolutely in some right half-plane independently of v, and the product

$$\prod_v \varsigma(s,W_v,f_v)$$

converges in this same half-plane.

As far as $W_{\varphi}(h)$ is concerned, we may choose $\varphi(g)$ so that $W_{\varphi}(g) = \Pi_v W_v(g_v)$, with each W_v non-zero in the local Whittaker model $\mathcal{W}(\pi_v,\psi_v)$. Indeed $\mathcal{W}(\pi,\psi) \neq \{0\}$ by assumption, and so the map $\varphi \to W_{\varphi}$ is an isomorphism onto $\Pi_v\mathcal{W}(\pi_v,\psi_v)$.

As far as $W_{\tau}^f(h,s)$ is concerned, let us choose once and for all a product definition of this function. Recall that $W_{\tau}^f(h,s)$ may be viewed as a complex-valued function on $H_{\mathbb{A}}$ such that $W_{\tau}^f(mh,s) = w(m)$ belongs to $\mathcal{W}(\tau\otimes \|^{s'+\frac{n-1}{2}},\psi^{-1})$ for each fixed h in $H_{\mathbb{A}}$. On the other hand, as an abstract representation space, $ind_{P_{\mathbb{A}}}^{H_{\mathbb{A}}}\tau\otimes\|^{s'} \approx \otimes_v\ (ind_{P_v}^{H_v}\tau\otimes\|_v^{s'})$, if $\tau = \otimes\tau_v$. Now fix f so that as a vector in the abstract space $\otimes ind\ \tau_v\otimes\|_v^{s'}$, f is of the form $\otimes f_v$, with each f_v in $ind\ \tau_v\otimes\|_v^{s'}$. Then $W_{\tau}^f(h,s)$ (regarded as a complex valued function) will be of the form $\Pi W_v^f(h_v,s)$, with each W_v^f in $ind\ \tau_v\otimes\|^{s'}$ such that $W_v^f(m_vh_v,s) = w_v(m)$ belongs to $\mathcal{W}(\tau_v\otimes\|^{s'+\frac{n-1}{2}},\psi_v)$, $s' = s-1/2$. The underlying

fact here is (a restricted infinite tensor product version of) the following: suppose τ_i is contained in $ind_{Z_i}^{M_i}\psi_i$ for $i = 1, 2$, and $f = f_1 \otimes f_2$ as a vector in the abstract space $(ind_{P_1}^{H_1}\tau_1) \otimes (ind_{P_2}^{H_2}\tau_2) \approx ind_{P_1 \times P_2}^{H_1 \times H_2}\tau_1 \otimes \tau_2$;
then regarding f concretely as a scalar valued function on $H_1 \times H_2$ gives
$f(h) = f_1(h_1)f_2(h_2)$.

(5.2) Proposition. Given $W_\varphi = \Pi W_v$, and $W_\tau^f = \Pi W_v^f$, there exists a positive real number s_0 such that each of the integrals $\varsigma(s, W_v, f_v)$ converges absolutely for $Re(s) > s_0$ (independently of v).

Proof. We give details of the proof only for the groups $G = SO(2n+1)$; minor modifications make the proof work for the other examples as well.

Because the functions W_v and f_v belong to admissible representation spaces, each is invariant by an open compact subgroup of the maximal compact subgroup of H. By Iwasawa's decomposition, we are therefore reduced to analyzing a finite sum of integrals of the form

$$\int_A W_v(a) W_{\tau_v}(a) |det\ a|^{s^*} d^*a$$

where A denotes the diagonal subgroup of H (realizable also as the diagonal subgroup of $GL(n)$), W_{τ_v} belongs to the Whittaker space $\mathcal{W}(\tau_v, \psi^{-1})$ on $GL(n)$, and s^* is a translate of s which does not depend on v.

Now we recall some elementary facts from the general theory of Whittaker functions. According to Proposition 6.1 and Lemma 5.1 of [Casselman-Shalika], there exists a positive number ε_v such that $W_v(a)W_{\tau_v}(a) = 0$ unless

$$a = \begin{pmatrix} a_1 & & & \\ & \ddots & & \\ & & \ddots & \\ & & & a_n \end{pmatrix}$$

satisfies $|a_i| \le \varepsilon_v$ for all i. In fact, for those W_v and W_{τ_v} which are "unramified", ε_v may be taken equal to 1. (We could not have concluded that $|a_i| \le \varepsilon_v$ by considering W_{τ_v} alone; indeed it is the root of G lying "outside $GL(n)$" which forces the inequality $|a_i| \le \varepsilon_v$.)

In any case, the conclusion is that the integrand $W_v W_{\tau_v}(a_i, \cdots, a_n)|det\ a|^{s^*}$ is compactly supported on k_v^n.

Next we recall that Whittaker functions attached to unitary representations are bounded functions. Indeed, from the arguments used in §1 of [Howe], it follows that such Whittaker functions are positive definite functions. This, together with the results of the last two paragraphs, implies that the integrals defining $\varsigma(s, W_v, f_v)$ converge for $Re(s)$ greater than some approximately chosen $s_0 > 0$.

In Section 13 we shall normalize our Eisenstein series $E(h, s)$ so that the resulting local zeta integrals $\varsigma(s, W_v, f_v)$ coincide with the expected local Langlands factors $L(s, \pi_v, r_v)$. From this discussion it will follow (from the general theory of Langlands; cf. §13 of [Borel]) that the product of these local zeta-integrals indeed converges. For more definitive results on the half-plane of convergence in the case of $GL(n) \times GL(n)$, cf. Sections 2.5 and 5.1 of [JS].

(5.3) The arguments above lead immediately to the following:

Corollary 5.3. For $Re(s)$ sufficiently large, $I(s, \varphi, E)$ factors as the product of local zeta-integrals

$$\varsigma(s, W_v, f_v) = \int_{U_v^H \backslash H_v} W_v(h) W_{\tau_v}^f(h, s) dh$$

where W_v is in $\mathcal{W}(\pi_v, \psi_v)$ and $W_{\tau_v}^f$ is such that for each h_v in H_v, the function $W_{\tau_v}^f(m_v h_v, s)$ belongs to $\mathcal{W}(\tau_v \otimes ||_v^{s^*}, \psi_v^{-1})$.

Additional Remarks.

For groups of type C_n, the situation is exactly the same, except for the presence of the function $F(g) = r(g)\Phi(e_n)$ in the integrand (compensating for the fact that the subgroup H fails to appear). In this case, F still factors as $\Pi\ F_v$, since the Schwartz-Bruhat function Φ may be taken to be a product of local functions $\Pi\Phi_v$. Thus we led to the Euler product

$$\prod \varsigma(s, W_v, \phi_v, f_v)$$

with

$$(5.3.1) \qquad \varsigma(s, W_v, \phi_v, f_r) = \int_{U_v^H \backslash H_v} W_v(g) F_v^{\phi_v}(g) W_{\tau_v}^f(g, s) dg.$$

For groups of type D_n, the situation is slightly more complicated; indeed the parabolic subgroup P used to define the induced functions $W^\tau(h, s)$ in (3.3.4) is not compatible with the Iwasawa decomposition for $G \subset H$. Of course the global integral (on the right hand side of (3.3.4)) still leads to a product of local integrals of the form

$$(5.3.2) \qquad \int_{\mathsf{U}_v^G \backslash G_v} W_\pi(g) W_\tau(g, s) dg.$$

However, if we write the Iwasawa decomposition for G_v as $\mathsf{U}^G T^G K_G$, where K_G is a maximal compact subgroup, then the convergence of (5.3.2) is reduced to analyzing local integrals of the form

$$(5.3.3) \qquad \int_{T^G} W_{\pi_v}(a) W_{\tau_v}(i(a)) |det\ (i(a))|^{s^*} d^* a,$$

where $i(a)$ denotes the projection from H to the maximal torus of the Levi component of P.

§6. Concluding Remarks.

At the beginning of Section 1, we remarked that Methods A, B and C can often be applied to the same group. In particular, we mentioned that the application of Methods A and B to the group $G = SO(2n+1)$ yields L-functions for $G \times GL(n+1)$ as well as $G \times GL(n)$. Similarly, both these methods can be applied to the group $G = SO(2n)$ of type D_n, the result being a theory of L-functions for $G \times GL(n-1)$ as well as $G \times GL(n)$.

We note that in the application of Method A to the group $G = SO(2n) \supset H = SO(2n-1)$, the required subgroup Q in axiom (1.1.1) is not a parabolic subgroup. Indeed, suppose $G = SO(V)$, where $V = X \oplus \{e_1\} \oplus \{e_2\} \oplus X^v$, X and X^v are (dual) isotropic subspaces of dimension $n-1$, and $(e_i, e_i) = 1$. Then we take $V' = X \oplus \{e_1\} \oplus X^v$, $H = SO(V') \subset G$, and P the maximal parabolic subgroup of H stabilizing X. If $\tilde{Q}$ is the parabolic subgroup of G preserving X, then $\tilde{Q} \approx GL(n-1)SO(2)\tilde{U}$, and we take $Q \subset \tilde{Q}$ to be the subgroup isomorphic to $GL(n-1)\tilde{U}$. With this choice of $G \supset H$, $Q \supset P$, it can be checked that Axioms (1.1.1) and (1.1.2) hold (with $n-1$ in place of n).

Some other classes of groups we can apply Method A to are the spin groups covering $SO(2n+1)$ or $SO(2n)$, and the quasi-split unitary groups $U_{n+1,n}$. For the latter groups, we proceed as for SO_{2n+1}, with P and Q (essentially) the parabolic subgroups of G and $H = U(V')$ preserving the maximal isotropic subspace X of V (or V'). For the case $n = 2$, this work was carried out in detail in [GePS] in order to lift π on $U_{2,1}$ to Π on $GL(3, K)$.

Finally, we note that the familiar Rankin-Selberg method for studying L-functions on $GL(n) \times GL(n)$ or $GL(n)$ over a quadratic extension resembles - but is not really the same as - our Method A. For example, let $G = GL(n) \times GL(n)$, and take $H = GL(n)$ to be diagonally imbedded in G. In [JS], L-functions on G are studied via the Rankin-Selberg integral

$$\int_{Z_{\mathbb{A}} H_k \backslash H_{\mathbb{A}}} \varphi(g)\varphi'(g)E(g, \Phi, s, \eta)dg$$

where φ and φ' belong to automorphic cuspidal representations Π and Π' of $GL_n(\mathbb{A})$, and $E(g, \Phi, s, \eta)$ is an Eisenstein series on $GL_n(\mathbb{A})$ induced (roughly) from the character $||^s\eta$ on the Borel subgroup of $GL_n(\mathbb{A})$.

As far as Method C is concerned, we remark that it can also be applied to the class of unitary groups $U_{n,n}$, as well as the orthogonal groups $SO(2n)$. In fact, this was the original way we approached the theory of L-functions on $G \times GL(n)$ for these groups.

We close this section by explaining how Method C may be viewed (at least heuristically) as being "dual" to Method A. Fix $G = Sp(n)$, with L-group ${}^LG = SO(2n+1)$, and subgroup $SO(2n) = {}^LH$. On the one hand, $SO(2n)$ and $SO(2n+1)$ are the groups which comprise the principal data for Method A, where restriction to ${}^LH$ plays the crucial role. On the other hand, at least over finite fields, the subgroup $SO(2n)$ of $SO(2n+1)$ determines a "small" representation of G via Lusztig's theory; indeed $SO(2n)$ is the centralizer of a semi-simple element g_0 in $SO(2n+1)$, and the resulting conjugacy class in ${}^LG$ corresponds to a "small" representation of G. If we view the Weil representation r of G as the local field analogue of this small representation, then we may view r as playing the same role in Method C (for $G = Sp(n)$) as H plays in Method A (for ${}^LG = SO(2n+1)$).

Chapter II : The Functional Equation

Throughout this Chapter, we shall assume that k is a local non-archimedean field of residual characteristic q, and τ is an irreducible admissible representation of $GL_n(k)$ with Whittaker model $\mathcal{W}(\tau, \psi^{-1})$. In Chapter I, our basic identity led us to consider "local" zeta-integrals of the form

$$(I) \qquad \varsigma(s, W, f^{\tau}) = \int_{U^H \backslash H} W(h) f(h, s) dh$$

where W belongs to the Whittaker model of an irreducible admissible representation π of G, and f belongs to a certain parabolically induced representation of $H \subset G$. Recall that the exact form of the integral defining ς depends on whether G is of type B_n, C_n or D_n ; for example, for $G = Sp(2n)$, the definition of ς must also involve the local component of a theta-series.

Our goal in this Chapter is to prove a functional equation for the zeta-integral $\varsigma(s, W, f)$. In particular, we shall prove that ς - initially defined only in some right half-plane by a convergent integral - is actually a rational function in q^{-s} , hence meromorphic in all of $\mathbb{C}$. More significantly, we shall prove that there exists a rational function $\gamma(s, \pi \times \tau, \psi)$ - independent of the choice of W and f, such that

$$(II) \qquad \varsigma(1 - s, W, M(s)f) = \gamma(s, \pi \times \tau, \psi)\varsigma(s, W, f).$$

Here $M(s)$ is the intertwining operator between $Ind\ \tau \otimes |det|^{s'}$ and $Ind\ \tilde{\tau} \otimes ||^{-s'}$ corresponding to the longest element of the appropriate Weyl group. We recall that $s' = s - 1/2$; therefore our functional equation (II) really involves the familiar substitution $s' \to -s'$.

As in Chapter I, we shall require distinct arguments for the cases B_n, C_n , and D_n. In each case, however, the functional equation shall result from proving the uniqueness of invariant bilinear forms on $V_{\pi} \times V_{Ind\ \tau||^s}$. (For $Sp(2n)$ we need to investigate a trilinear form.) Also common to these three cases is the fact that the rationality (and hence the meromorphic continuation) of $\varsigma(s)$ will in fact follow from the uniqueness of these H-invariant forms; cf. Section 12.

§7. Groups of Type B_n : Preliminaries and Statement of the Main Theorem

(7.1) Here $G = SO(V) = SO(2n+1)$, $H \approx SO(V') \subset G$, and π is an irreducible admissible nondegenerate representation of G . Regard the Whittaker model $\mathcal{W}(\pi, \psi)$ as an abstract representation space V_π for π. Then for each s in some right half-plane, the formula

$$\varsigma(s)(W, f) = \int_{U^H \backslash H} W(h) f^\tau(h, s) dh$$

defines a bilinear form $\varsigma(s)$ on $V_\pi \times V_{Ind\ \tau \otimes ||^{s'}}$. From the integral representation for $\varsigma(s)$, it is clear that this bilinear form is H-invariant, i.e., $\varsigma(s)(h \cdot W, h \cdot f^\tau) = \varsigma(s)(W, f^\tau)$ for all $h \in H$. Here we are regarding V_π as an H-module by restriction, and the action of h on either $\mathcal{W}(\pi, \psi)$ or $V_{Ind\ \tau \otimes ||^{s'}}$ is given by right translation.

Let us assume for the moment that our zeta-integral $\varsigma(s, W, f^\tau)$ is at least a meromorphic function in all of $\mathbb{C}$. Then we may also consider the bilinear form $\varsigma^*(s)$ defined on $V_\pi \times V_{Ind\ \tau \otimes ||^s}$ by $\varsigma^*(W, f^\tau) = \varsigma(-s', W, M(s)f)$, with $M(s)$ the intertwining operator between $Ind\ \tau \otimes |det|^{s'}$ and $Ind\ \tilde{\tau} \otimes |det|^{-s'}$ corresponding to the Weyl group element

$$W = \begin{bmatrix} 0 & -I \\ I & 0 \end{bmatrix}$$

of H. (Here $\tilde{\tau}$ denotes the representation contragredient to τ , with Whittaker model $\mathcal{W}(\tilde{\tau}, \psi^{-1}) = \{W({}^t x^{-1}) : W \in \mathcal{W}(\tau, \psi^{-1})\}$.) Because the operator $M(s)$ commutes with the action of H, the form $\varsigma^*(s)$ is also H-invariant.

Now recall that for any pair of smooth representations π_1, π_2 of a p-adic group H, the vector space of H-invariant forms on $V_{\pi_1} \times V_{\pi_2}$ is isomorphic to $Hom(\pi_1, \tilde{\pi}_2)$ or $Hom(\pi_2, \tilde{\pi}_1)$; cf. [BZ], §3.6. Therefore, to prove the required functional equation (II), it will suffice to prove:

Theorem 7.1. Outside of a finite set of values for q^s , the space $Hom_H(\pi, Ind_P^H\ \tau \otimes ||^s)$ is one-dimensional.

Indeed, this Theorem immediately implies the existence of a scalar factor $\gamma(s, \pi \times \tau, \psi)$ such that (II) holds; if, moreover, our zeta-integrals are known to be rational functions of q^{-s}, then also $\gamma(s, \pi \times \tau, \psi)$ will be rational.

The proof of Theorem 7.1 is a bit complicated. To simplify the exposition, we first collect some preliminary results on the representation theory of GL_n and its subgroups P_m ; then we present the proof of the Main Theorem in a sequence of several steps.

(7.2) Review of the theory of Bernstein-Zelevinsky

Recall from Proposition 2.2 of Chapter I that

$$U^P \setminus Q \approx P_{n+1}.$$

Here Q is the maximal parabolic subgroup of G preserving X , and U^P is the unipotent radical of the parabolic $P = Q \cap H$ of H. Of course $P_{n+1} = \left\{p = \left[\begin{smallmatrix} g & * \\ 0 & 1\end{smallmatrix}\right] \in GL_{n+1} : g \in GL_n\right\}$. The purpose of this paragraph is to recall the basic structure of an arbitrary P_{n+1} module.

For each $0 \leq i \leq n$, denote by R_i the subgroup of P_{n+1} of the form

$$R_i = \left\{ \begin{bmatrix} \overbrace{g}^{i} & \overbrace{v}^{n+1-i} \\ 0 & z \end{bmatrix} \right\}$$

where $g \in GL_i(k)$, $v \in M_{i,n+1-i}$, and

$$z \in Z_{n+1-i} = \left\{ \begin{bmatrix} 1 & & * \\ & \ddots & \\ & & 1 \end{bmatrix} \in GL_{n+1-i}(k) \right\}.$$

For example, $R_0 = Z_{n+1} \subset P_{n+1}$, and $R_n = P_{n+1}$. The unipotent radical of $R_n = P_{n+1}$ is

$$V_n = \{g_{ij} \in P_{n+1} : g_{ij} = \delta_{ij} \;\; for \;\; j < n+1\}$$

and P_{n+1} is the semidirect product of GL_n (embedded in the obvious way) with V_n .

Define a character θ of the subgroup V_n by the formula

$$\theta((v_{ij})) = \psi(v_{n,n+1}).$$

Using this character of V_n, we want to define a (modified) Jacquet functor taking representations of P_{n+1} to representations of $P_n \subset P_{n+1}$. Consider the subgroup $S_n = P_n \times V_n$ of $P_{n+1} = GL_n \times V_n$; this is precisely the subgroup of P_{n+1} which stabilizes the character θ of V_n . Given a smooth representation σ of P_{n+1} , we define a subspace $E_{\sigma,\theta}$ of the space E_σ of σ through the formula

$$E_{\sigma,\theta} = \{\sigma(u)\xi - \theta(u)\xi : u \in V_n,\ \xi \in E_\sigma\}.$$

Because V_n is normal in P_{n+1} , and $P_n \times V_n$ stabilizes θ, the subspace $E_{\sigma,\theta}$ is invariant for $S_n = P_n \times V_n$.

Denote by $E_\sigma(\theta)$ the resulting quotient of E_σ by $E_{\sigma,\theta}$. The representation $\sigma|_{S_n}$ then passes to this quotient and defines a representation σ_θ , through which V_n acts according to scalar multiplication by θ . By $r_{V_{n,\theta}}(\sigma)$ we denote the restriction of σ_θ to P_n, and we refer to the functor

$$\sigma \longrightarrow r_{V_{n,\theta}}(\sigma)$$

as the modified Jacquet functor from P_{n+1} to P_n .

Remark. The "ordinary Jacquet functor" corresponds here to the case $\theta \equiv 1$; in this case, the stabilizer S_n is all of $P_{n+1} = GL_n \times V_n$ and the functor $\sigma \rightarrow r_{V_{n,1}}(\sigma)$ goes from P_{n+1} to GL_n.

Now we need to recall the notion of "derivative" for representations of P_{n+1} . The easiest way to do this is via the following diagram:

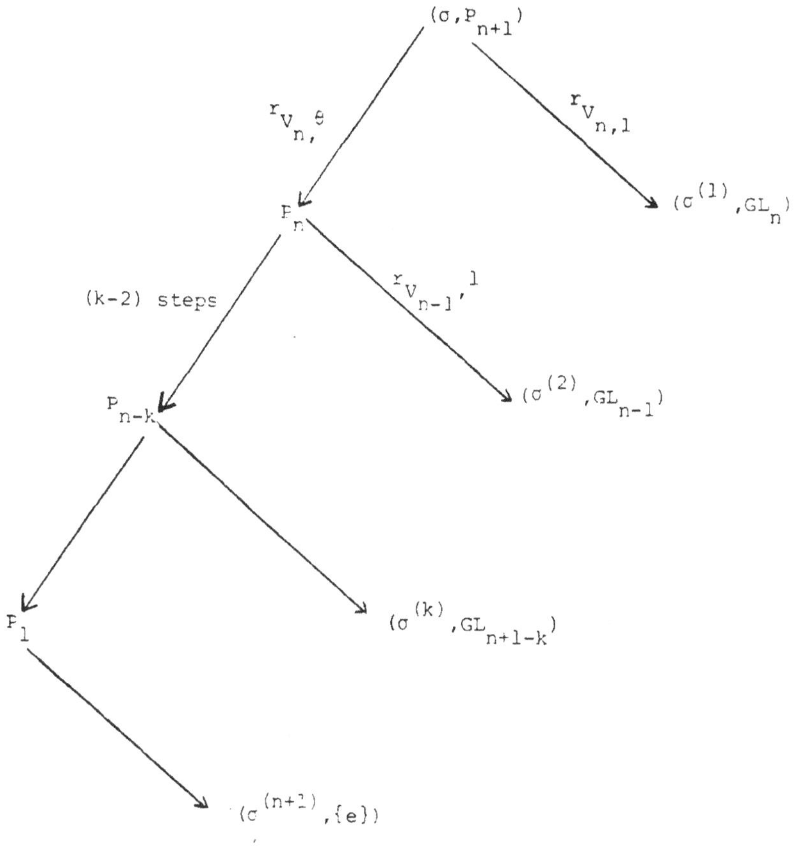

In this diagram, the topmost vertex corresponds to our given smooth representation σ of P_{n+1}. The rightward (resp. leftward) pointing arrows refer to the appropriate "ordinary" (resp. "modified") Jacquet functors from P_{i+1} to GL_i (resp. P_i). The k-th derivative of σ is then a representation of GL_{n+1-k} obtained from σ by first proceeding $k-1$ steps "to the left", and then once to the right, i.e.,

$$\sigma^{(k)} = r_{V,1} \circ (r_{V,\theta})^{k-1}(\sigma).$$

For details on this and the preceding constructions, see §3 of [BZ]; the idea of relating the representations of P_i and GL_n goes back to Gelfand and Kazhdan ([GelKaz]). Gelfand and

Kazhdan also proved the following: if σ is the restriction to P_n of an irreducible cuspidal representation of GL_n, then $\sigma^{(k)} = 0$ for $0 < k < n$, and $\sigma^{(n)} = 1$ is one-dimensional. (In general, by the derivative of a representation of GL_n we understand the derivative of its restriction to P_n.)

Finally, we recall the following basic:

Structure Theorem 7.2. ([BZ], §3.) Suppose (σ, E_σ) is an arbitrary smooth P_{n+1} module. Then there exist invariant subspaces

$$0 = W_0 \subset W_1 \subset \cdots \subset W_{n+1} = E_\sigma$$

such that (as P_{n+1} - modules)

$$W_{i-1} \setminus W_i \approx ind_{R_{i-1}}^{P_{n+1}} \sigma^{(n-i+2)} \otimes \psi$$

for each $i = 1, \cdots, n+1$.

Here ψ is regarded as a character of $Z_{n-i+2} \subset R_{i-1}$ via the formula $\psi((z_{ij})) = \psi(\sum_{j+1}^{n-i+1} z_{j,j+1})$, and $\sigma^{(n-i+2)} \otimes \psi$ is extended from $GL_{i-1} \times Z_{n-i+2}$ to R_{i-1} trivially across

$$N_{i-1} = \left\{ \begin{bmatrix} r_{i-1} & v \\ 0 & I_{n+2-i} \end{bmatrix} \right\} .$$

Note that when $i = 1$, $\sigma^{(n-i+2)}$ is the $(n+1)$-th derivative of σ and this - being a representation of $\{e\}$, must be a multiple of the identity. Moreover, in this case, $R_0 = Z_{n+1}$. Therefore W_1 is equivalent to a multiple of $ind_{Z_{n+1}}^{P_{n+1}} \psi$, the famous (irreducible) "standard representation" of Gelfand-Graev; cf. [GelKaz].

§8. Groups of Type B_n : Reduction of the Main Theorem

Our goal is to prove that (outside of a finite set of values for q^s) $Hom_H(\pi, Ind_P^H \tau \otimes ||^s)$ is one-dimensional.

(8.1) First Reduction. Observe that

(8.1.1) $$Hom_H(\pi, ind_P^H \tau \otimes ||^s) \cong Hom_M(V_\pi^{U^P}, \tau \otimes |det|^s),$$

where M (resp. U^P) is the Levi component (resp. unipotent radical) of the parabolic P, and $V_\pi^{U^P}$ is the Jacquet module $V_\pi / V_\pi(U^P)$, with $V_\pi(U^P) = \{v \in V_\pi : \int_{N_0} \pi(n) v dn = 0 \; for\ some\ open\ compact\ subgroup\ N_0 \text{ of } U^P\}$.

Proof. Apply the Frobenius reciprocity theorem (cf. Theorem 3.2.5 of [Casselman]) to the smooth representation $\pi|_H$ of H. The subspace $V_\pi^{U^P}$ may also be described in terms of the generators $\{\pi(n)v - v : \; n \text{ in } U^P, \text{ v in } V_\pi\}$; cf. [BZ2], §2.35.

Caution. Because U^P is the unipotent radical of a parabolic of H (as opposed to G) , $V_\pi^{U^P}$ is not a bona fide Jacquet module for V_π . Nevertheless, it is still an M-module, and it is this module whose Jordan-Holder series plays a key role in the proof of Theorem 7.1.

To analyze $V_\pi^{U^P}$ as a GL_n - module, it turns out to be useful to first analyze it as a P_{n+1} - module, and then restrict matters to GL_n . (This is feasible, since $P_{n+1} \approx Q/U^P$, and the definition of V_π^P is such that any subgroup of G which - like Q - normalizes U^P also acts naturally in $V_\pi^{U^P}$.)

(8.2) Second Reduction : We claim Theorem 7.1 is a corollary of the following:

<u>**Proposition 8.2.**</u> There exists a finite sequence of P_{n+1} - submodules

$$\{0\} = V_0 \subset V_1 \subset \cdots \subset V_N = V_\pi^{U^P}$$

such that each $V_{i-1} \setminus V_i$ is an <u>irreducible</u> P_{n+1} module, $V_0 \setminus V_1 \cong ind_{Z_{n+1}}^{P_{n+1}} \psi$, and $V_1 \setminus V_\pi^{U^P}$ never contains $ind_{Z_{n+1}}^{P_{n+1}} \psi$ as a subquotient.

By the first reduction, it suffices to show that $Hom_{GL_n}(V_\pi^{U^P}, \tau \otimes ||^s)$ has dimension 1 (outside of a finite set of values for q^s). Assuming Proposition 8.2 is true, it remains to prove that

$$(a) \qquad Hom_{GL_n}(Res_{GL_n}^{P_{n+1}} ind_{Z_{n+1}}^{P_{n+1}} \psi, \tau \oplus ||^s) = 1$$

for all s, and

$$(b) \qquad outside\ of\ a\ finite\ set\ of\ values\ for\ q^s,$$

$$Hom_{GL_n}(Res_{GL_n}^{P_{n+1}} V_1 \setminus V_\pi^{U^P}, \tau \otimes ||^s) = 0.$$

As far as (a) is concerned, we note that $P_{n+1} = GL_n Z_{n+1}$ implies that there is only one double coset of P_{n+1} with respect to GL_n and Z_{n+1}. Therefore, by Mackey's restriction theorem

$$Res_{GL_n}^{P_{n+1}}\ ind_{Z_{n+1}}^{P_{n+1}} \psi = ind_{Z_n}^{GL_n} \psi$$

since $GL_n \cap Z_{n+1} \approx Z_n$. In other words,

$$Hom_{GL_n}(Res_{GL_n}^{P_{n+1}}\ ind_{Z_{n+1}}^{P_{n+1}} \psi, \tau \otimes ||^s)$$

$$= Hom_{GL_n}(ind_{Z_n}^{GL_n} \psi, \tau \otimes ||^s).$$

But by assumption, $\tau \otimes ||^s$ is non-degenerate. This implies, on the one hand, that this last *Hom* is non-empty, and on the other hand - by uniqueness of Whittaker models, that the dimension of this *Hom* space is less than or equal to 1. This proves (a).

To prove (b), we shall again apply Mackey's theorem. First we note that any <u>irreducible</u> P_{n+1}-module σ must be of the form

$$(8.2.1) \qquad ind_{R_m}^{P_{n+1}} \omega \otimes \psi$$

where $0 \leq m \leq n$, ω is an irreducible admissible representation of $GL_m(k)$, and both m and ω are uniquely determined by σ. This is an immediate consequence of the Structure Theorem for arbitrary P_{n+1}-modules (cf. Cor. 3.5 of [BZ]). Applying Mackey's restriction theorem to σ, we have

$$Res_{GL_n}^{P_{n+1}}\ ind_{R_m}^{P_{n+1}}\omega\otimes\psi = ind_{GL_mZ_{n-m}N_m^*}^{GL_n}\omega\otimes\psi,$$

since $P_{n+1} = GL_nR_m$ and $GL_n\cap R_m \approx GL_mZ_{n-m}N_m^*$, with $N_m^* \approx M_{m,n-m}(k)$. Moreover, by Proposition 8.2, we know that any irreducible constituent of $V_1\setminus V_\pi^{U^P}$ must have parameters (m,ω), with $m\geq 1$. So to prove (b), it remains to prove that - for $m\leq 1$,

$$Hom_{GL_n}(ind_{GL_mZ_{n-m}N_m^*}^{GL_n}\omega\otimes\psi,\ \tau\otimes\|^s) = 0 \tag{8.2.2}$$

outside of a finite set of values for q^s.

For this, we need to bring into play some basic facts concerning an arbitrary irreducible admissible representation of $GL_n(k)$ like τ. From Theorem 2.5 of [BZ], it follows that such a τ is a subrepresentation of some $ind_{P_{n_1,\cdots,n_r}}^{GL_n}\tau_1\otimes\cdots\otimes\tau_r$ with $P_{n_1,\cdots,n_r}$ the standard parabolic subgroup of $GL_n(k)$ with Levi component $\Pi_{i=1}^rGL_{n_i}$, and each τ_i an irreducible cuspidal representation of $GL_{n_i}(k)$. Moreover, this "cuspidal data" $(\tau_1,\cdots,\tau_r,\ P_{n_1},\cdots,n_r)$ is uniquely determined (up to ordering) by τ. To prove (8.2.2), it therefore suffices to prove a similar statement for

$$Hom_{GL_n}(ind_{GL_mZ_{n-m}N_m^*}^{GL_n}\omega\otimes\psi,\ ind_{P_{n_1,\cdots,n_r}}^{GL_n}\tau_1\|^s\otimes\cdots\otimes\tau_r\|^s)$$

By induction in stages, i.e., first from GL_mZ_{n-m} to GL_mGL_{n-m} - where we already obtain all possible irreducible non-degenerate representations β of GL_{n-m}, it even suffices to analyze

$$Hom_{GL_n}(ind_{P_{n'_1,\cdots,n'_\ell}}^{GL_n}\omega_1\otimes\cdots\otimes\omega_\ell,\ ind_{P_{n_1,\cdots,n_r}}^{GL_n}\tau_1\|^s\otimes\cdots,\otimes\tau_m\|^s). \tag{**}$$

Recall, finally, that the representations τ_i are uniquely determined by τ. Since we can assume $m\geq 1$, at least 1 of these ω_i's must similarly be determined by ω, call it ω_{i_0}. Therefore, by another well-known theorem of Bernstein and Zelevinsky ([BZ], Theorem 2.9), our Hom space (**) can be non-zero only if at least one of the $\tau_i|det|^s$ is equivalent to ω_{i_0} . In other words, since τ (and hence each τ_i) is fixed from the start, there is only one possible value of $||^s$ (hence of q^s) which can make this Hom space non-zero. On the other hand, Proposition 8.2 implies that there are only a finite number of possibilities for such ω on GL_m, with $m \geq 1$; outside of the finite number of resulting values for q^s, we must therefore have

$$Hom(V_1 \setminus V_\pi^{U^P}, \tau \otimes ||^s) = 0,$$

as claimed in (b).

§9. Groups of Type B_n : Proof of the Main Theorem (and Functional Equation)

From the reductions of §8 it remains to prove Proposition 8.2 describing the Jordan-Holder series for $V_\pi^{U^P}$ (regarded as a P_{n+1}-module). The first step is to examine the "normal" series provided for $V_\pi^{U^P}$ by the General Structure Theorem of §7.2.

Step 1. Suppose

$$\{0\} = W_0 \subset W_1 \subset \cdots \subset W_{n+1} = V_\pi^{U^P}$$

is the normal series provided by Theorem 7.2. Then we claim that W_1 is (irreducible and) isomorphic to $ind_{Z_{n+1}}^{P_{n+1}}\psi$, but that $ind_{Z_{n+1}}^{P_{n+1}}\psi$ never occurs (as a subquotient) of $W_1 \setminus V_\pi^{U^P}$.

Proof. First we shall show directly that $ind_{Z_{n+1}}^{P_{n+1}}\psi$ must occur at least once in $V_\pi^{U^P}$, i.e., we shall show that there exists subspaces invariant $W_1^* \supset W_2^*$ of $V_\pi^{U^P}$ such that $W_1^*/W_2^* \approx ind_{Z_{n+1}}^{P_{n+1}}\psi$. By assumption on π, we know at least that V_π itself admits a non-trivial Whittaker functional, i.e., there exists a non-zero functional $\ell : V_\pi \rightarrow \mathbb{C}$ such that $\ell(\pi(u)v) = \psi(u)\ell(v)$ for all u in $U^G = Z_{n+1}U^Q$. Clearly ℓ is trivial on the subspace

$$V_\pi(U^P) = \{\pi(u)v - v : u \in U^P, v \in V_\pi\}$$

since ψ is trivial on U^P. Therefore ℓ is defined (and non-zero) on $V_\pi^{U^P}$.

Define a P_{n+1}-module map

$$L : V_\pi^{U^P} \longrightarrow Ind_{Z_{n+1}}^{P_{n+1}}\psi$$

by $L(v)(p) = \ell(\pi(p)\cdot v)$. Here we are following [BZ] in using "Ind" (as opposed to "ind") to denote "ordinary" (as opposed to "compact") induction. Also, as always, p in P_{n+1} is acting on v in $V_\pi^{U^P}$ thru its pullback in Q (recall the projection $\alpha : Q \rightarrow P_{n+1}$ of Proposition (2.2)). Now let

$$W_2^* = \{v \in V_\pi^{U^P} : L(v) \equiv 0\}$$

and

$$W_1^* = \{v \in V_\pi^{U^P} : L(v) \in ind\ \psi\},$$

i.e. $L(v)(p)$ *is compactly supported modulo* Z_{n+1}.

Then $W_2^* \subset W_1^*$, and the image of W_1^* under L is all of $ind\ \psi$ since this latter representation is irreducible. Thus $W_1^*/W_2^* \approx ind_{Z_{n+1}}^{P_{n+1}}\psi$, and so $ind\ \psi$ occurs at least once in $V_\pi^{U^P}$.

Next suppose we are given some pair of invariant subspaces

$$W' \subset W \subset V_\pi^{U^P}$$

with the property that $W' \setminus W \approx ind_{Z_{n+1}}^{P_{n+1}}\psi$. We claim then that no $W'' \not\subset W'$ contains $ind\ \psi$ as a factor module.

To see this, consider the natural "Whittaker" functional ℓ on the space of $ind_{Z_{n+1}}^{P_{n+1}}\psi$ defined by $\ell : f(p) \mapsto f(1)$; it is non-zero and such that

$$\ell(\pi(z)f) = \psi(z)\ell(f) \quad \textit{for all} \quad z \in Z_{n+1}.$$

Therefore, pulling ℓ up to W via the isomorphism $ind\ \psi \approx W/W'$, we get a non-trivial functional $\ell : W \rightarrow \mathbb{C}$ such that

$$\ell(\pi(u)w) = \psi(u)\ell(w)$$

for all $w \in W$ and all u in the pullback of Z_{n+1} in Q. Clearly this pullback is U^G, the unipotent radical of G. In order to further pullback ℓ to a genuine Whittaker functional on V_π, we appeal to the following:

__Lemma 9.1.__ Suppose N is a unipotent p-adic group with character ψ, and $W_1 \subset W_2 \subset V$ are N-modules. If Y is a space on which N acts according to ψ, then each linear N-map $\ell : W_1 \rightarrow Y$ extends to an N-map ℓ on W_2, i.e., $\ell(n \cdot w') = \psi(n)\ell(w')$.

This follows immediately from the exactness of the Jacquet functors, more precisely, Prop. 2.35 of [BZ]. In the present context, we use it to extend our $\ell : W \rightarrow \mathbb{C}$ up to $V_\pi^{U^P}$; then, through the natural projection $V_\pi \rightarrow V_\pi^{U^P}$, we further pull ℓ back to V_π itself. The result is a bona fide Whittaker functional on V_π which vanishes on the pullback of W in

V_π. Uniqueness of Whittaker models for V_π then implies that no $W'' \not\subseteq W'$ can contain *ind* ψ as a factor module, as claimed.

Getting back to the claim made in Step 1, consider again the normal series $0 = W_0 \subset W_1 \subset \cdots \subset W_{n+1} = V_\pi^{U^P}$ provided for by the basic Structure Theorem. Recall that W_1 is isomorphic to a multiple of $ind_{Z_{n+1}}^{P_{n+1}}\psi$. The analysis just given implies that this multiple must be at most one, and that no $W_1 \backslash W_i$, $i > 1$, can contain *ind* ψ as a quotient representation. On the other hand, the very first part of the analysis in Step 1 implies *ind* ψ must occur at least once. Hence the claim of Step 1 follows.

The rest of the proof of Proposition 8.2 will involve showing that $W_1 \backslash V_\pi^{U^P}$ has a finite Jordan Holder series. Indeed, this step (Step 2) will complete the proof of the Proposition. The basic plan is to bring into play some "genuine" Jacquet modules for π, i.e., Jacquet modules with respect to the unipotent radicals of parabolic subgroups of G itself. Although the constructions are a bit complicated, the idea is simple, and may be explained as follows.

Roughly speaking, each non-trivial constituent of $JH(W_1 \backslash V_\pi^{U^P})$ (the Jordan-Holder series for $W_1 \backslash V_\pi^{U^P}$) will give rise via these constructions to a distinct (non-trivial) constituent of some genuine Jacquet module; the standard finiteness theorems for these latter modules will then in turn put sufficient restrictions on the possible size of $JH(W_1 \backslash V_\pi^{U^P})$. For example, suppose for the moment that π happens to be cuspidal. Then any genuine Jacquet module of π is trivial, and we shall see that $W_1 \backslash V_\pi^{U^P}$ cannot have any constituents at all, i.e. $V_\pi^{U^P}$ is actually irreducible (and hence equivalent to $ind_{Z_{n+1}}^{P_{n+1}}\psi$). For arbitrary V_π, the argument is similar but less transparent.

We now begin this task in earnest.

Step 2. Prove that $W_1 \backslash V_\pi^{U^P}$ has a finite Jordan-Holder series.

The proof proceeds via a sequence of Lemmas.

Lemma 9.2. For each $m \geq 1$, let Q_m denote the maximal parabolic subgroup of G which stabilizes the m-dimensional isotropic subspace $< x_1, \cdots, x_m >= X_m$ inside $X =< x_1, \cdots, x_n >$. Recall the projection map $\alpha : Q \to P_{n+1}$, and the subgroup

$N_m = \left\{ \begin{bmatrix} I_m & * \\ & I_{n+1-m} \end{bmatrix} \right\}$ of P_{n+1}. Then $\alpha^{-1}(N_m)$ contains the unipotent radical U_m of Q_m, and $Q_m/U_m \approx GL_m \times SO_{2m-2m+1}$.

Proof. Write $X = X_m + L_0 + X_m^v$, where $L_0 =< x_{m+1}, \cdots, x_n, \ell_0,$ $x_n^v, x_{n-1}^v, \cdots, x_{m+1}^n >$ has dimension $2n - 2m + 1$, and the restriction of the bilinear form $(\ , \ )$ to L_0 remains non-degenerate. Arguing exactly as in the proof of Proposition 2.1, we define a map

$$t : Q_m \longrightarrow GL(X_m) \times \times SO(L_0)$$

which takes q to $(q|_{X_m}, q|_{(X_m+L_0)/X_m})$. The unipotent radical U_m of Q_m is precisely the same thing as *ker* t, and α maps U_m to N_m, i.e., $\alpha^{-1}(N_m) \supset U_m$. Thus the Lemma is established.

Next we exploit some basic facts about genuine Jacquet modules. For any reductive group G, with parabolic subgroup MN, let π be an irreducible admissible representation of G and V_π^N the corresponding Jacquet module.

In this general context (cf. [BZ], §5), the Jordan-Holder series of V_π^N is known to be finite (as an M-module), with length depending only on π and P. In fact, this same result is valid for more general Jacquet modules, where V_π^N is defined in terms of an arbitrary character ψ of N normalized by M, i.e., $V_\pi^N = V_\pi^{N,\psi} = V_\pi / V_\pi(N, \psi)$ with

$$V_\pi(N,\psi) = \{\pi(n)v - \psi(n)v : n \in N, v \in V_\pi\}.$$

A special Corollary of this result is the following:

Lemma 9.3. (a) As a Q_m/U_m module, $V_\pi^{U_m}$ has finite length;

(b) Regarding $V_\pi^{U_m}$ as a $GL_m \times SO_{2n-2m+1}$ module, consider the parabolic subgroup $GL_m \times B$ of $GL_m \times SO_{2(n-m)+1}$, where B is the standard Borel subgroup of $SO_{2n-2m+1}$. Let $[V_\pi^{U_m}]^{Z_{n-m+1}^*, \psi^*}$ denote the (generalized) Jacquet module of $V_\pi^{U_m}$ with respect to the standard maximal unipotent subgroup Z_{n-m+1}^* of $SO_{2(n-m)+1}$ and the standard nondegenerate character ψ^* belonging to ψ. (Note $Z_{n-m+1}^* \approx \alpha^{-1}(Z_{n-m+1})$, and ψ^* is the pullback of the standard non-degenerate character ψ on Z_{n-m+1}). Then $[V_\pi^{U_m}]^{Z_{n-m+1}^*, \psi^*}$ has finite length as a GL_m-module.

Recall that our task is to prove that $W_1 \setminus V_\pi^{U^P}$ has a finite length as a P_{n+1}-module. We know already that any irreducible constituent of $JH\left[W_1 \setminus V_\pi^{\cup^P}\right]$ must be of the form

$$\pi_m = ind_{R_m}^{P_{n+1}} \omega \otimes \psi$$

with $m \geq 1$, and ω an irreducible admissible representation of $GL_m(k)$.

Lemma 9.4. If $m \geq 1$, and π_m occurs in $JH\left[W_1 \setminus V_\pi^{U^P}\right]$, then

$$V_\pi^{U_m} \neq 0.$$

Proof. Suppose $0 \subset W' \subset W$ are non-trivial invariant subspaces of $V_\pi^{\cup^P}$ with the property that (the P_{n+1}-module) $W' \setminus W$ is isotropic to this π_m. Then we can define an R_m-map

$$L_m : W' \setminus W \longrightarrow V_{\omega \otimes \psi}$$

by $f \mapsto f(e)$. Recall that $V_\pi^{\cup^P} = V_\pi / V_\pi(\cup^P)$ and $W \subset V_\pi^{\cup^P}$. Thus we may first pullup L_m to a map from W, and then back to a non-trivial map from $\tilde{W} \subset V_\pi$ to $V_{\omega\otimes\psi}$. Note that the subgroup $N_m Z_{n-m+1}$ of $R_m \subset P_{n+1}$ acts on $V_{\omega\otimes\psi}$ by scalar multiplication by the character ψ. In particular, N_m acts trivially on $V_{\omega\otimes\psi}$.

Now by Lemma 9.2, $\alpha^{-1}(N_m) \supset U_m$. Therefore the map $L_m : \tilde{W} \to V_{\omega\otimes\psi}$ factors through a non-trivial map from $\tilde{W}^{\cup_m} \subset V_\pi^{U_m}$ to $V_{\omega\otimes\psi}$. In particular, $V_\pi^{\cup_m}$ is non-trivial, as was to be shown.

Since π cuspidal implies $V_\pi^{U_m} = 0$, this lemma gives us:

Corollary 9.5. If π is cuspidal, then π_m never occurs in $V_\pi^{U_P}$. In particular, $V_\pi^{U_P}$ is irreducible, equivalent to $ind_{Z_{n+1}}^{P_{n+1}} \psi$ as a P_{n+1} module, and

$$dim(V_\pi^{\cup_P}, \tau \otimes \|^s) = 1 \ \underline{for\ all} \ s.$$

We turn finally to the proof of Proposition 8.2 for arbitrary π. Consider the non-trivial map $\tilde{W}^{U_m} \longrightarrow V_{\omega\otimes\psi}$ just constructed in the proof of Lemma 9.4 (still assuming that some π_m occurs in $JH(V_\pi^{U^P})$). If for the same $m \geq 1$ there is another pair of invariant subspaces $W_1' \subset W_1 \not\subseteq W$ such that $W_1' \setminus W_1 \approx \pi_m'$, then we get a new map L_m' from $\tilde{W}_1^{U_m} \not\subseteq \tilde{W}^{U_m}$ to $V_{\omega' \otimes \psi}$. Note that $\alpha^{-1}(N_m)Z_{n-m+1}^*$ still acts in $V_{\omega' \otimes \psi}$ according to the character ψ (or ψ^*). Applying the Jacquet functor $r_{Z_{n-m+1}^*, \psi^*}$, we thus obtain distinct submodules $\left[\tilde{W}_1^{U_m}\right]^{Z^*, \psi^*} \subset \left[\tilde{W}^{U_m}\right]^{Z^*, \psi^*}$ for the Jacquet module $\left[V_\pi^{U_m}\right]^{Z_{n-m+1}^*, \psi^*}$. But by lemma 9.3, this latter module has finite length as a GL_m module. Thus, for each fixed $1 \leq m \leq n$, there can be only finitely many pairs of subspaces in any normal series for $W_1 \setminus V_\pi^{U^P}$ whose quotients realize some $\pi_m = ind_{R_m}^{P_{n+1}} \omega \otimes \psi$. This completes the proof of Proposition 8.2 and hence the functional equation as well.

Concluding Remark. Recall that we assumed from the outset that $\varsigma(s, w, f^\tau)$ defines a rational (or at least meromorphic) function in q^{-s}. This is necessary in order to make sense out of the functional equation, since $\varsigma(s, W, M(s)f)$ is initially defined by an integral convergent only in some left halfplane. However, the uniqueness of H-invariant bilinear forms on $V_\pi \times V_{Ind\ \tau \otimes ||}$. was established independently of this assumption, and - as already suggested before - actually implies this assumption. Following J. Bernstein, we shall explain these ideas in more generality at the end of this Chapter.

§10. Groups of Type D_n .

In this case, our local zeta integral takes the form

$$\varsigma(s, W, f^\tau) = \int_{U^G \backslash G} W^\psi(g) f^\tau(g, s) dh,$$

where W_φ^ψ belongs to the Whittaker model $\mathcal{W}(\pi, \psi)$ of an irreducible non-degenerate representation π of G, and f^τ is an element of $ind_P^H \tau \otimes |det|^{s'}$ restricted to G. As in Section 7, our goal here will be to prove that (outside of a finite set of values for q^{-s}) the space of G-invariant bilinear forms on $V_\pi \times V_{ind\ \tau \otimes ||^s}$ is one-dimensional. In particular, assuming the rationality of our local zeta-integrals, this will imply the requisite functional equation.

We begin with some preliminaries on the space $P \backslash H / G$. As in Section 3, we denote the three G-orbits of the space of maximal isotropic subspaces of V by E, $Eh_1 = X$, and Eh_2. Our first goal is to study the contribution from each of these orbits to the space $Hom_G(\pi, Res_G^H\ ind_P^H\ \tau \otimes ||^s)$.

(10.1) Throwing Away the Negligible Orbits.

We shall show that the "negligible orbits" Eh_1 and Eh_2 indeed contribute negligibly to $Hom_G(\pi, Res_G^H\ ind_P^H\ \tau \otimes ||^s)$.

Recall that $E' = E \cap (X \oplus X^v)$,

$$P^* = \{h \in : h(E') = E'\},$$

and

$$Q = \{h \in G : h(E) = E\} = P \cap G.$$

In proposition (3.3.2) we proved that

$$P^* \approx GL_{n-1} \times SO(1,1) \times U^{P^*},$$

and

$$Q \approx GL_{n-1} U^{P^*}.$$

Here U^{P^*} is the unipotent radical of P^* . By Remark 3.3.1, we can also write

$$Q \approx P_n \cup^Q \supset Z_n \cup^Q$$

where $Z_n \cup^Q$ is a maximal unipotent subgroup U^G of G.

Lemma 10.1.1. $Res_G^H \; ind_P^H \; \tau \otimes ||^s$ is the sum of three induced representations: $ind_Q^G \tau \otimes ||^s$, plus two induced representations belonging to the orbits of Eh_1 and Eh_2 .

Proof. By Mackey's restriction theorem,

$$\begin{aligned} Res_G^H \; ind_P^H \; \tau \otimes ||^s = & \; ind_Q^G \; \tau \otimes ||^s \\ & \oplus ind_{Q^{h_1}} \tau^{h_1} \otimes ||^s \\ & \oplus ind_{Q^{h_2}} \tau^{h_2} \otimes ||^s \end{aligned}$$

where $Q^h = h^{-1} P h \cap G$, and τ^h denotes the representation of Q^h defined by $\tau^h(g) = \tau(hgh^{-1})$.

The fact that each $Q^{h_i} = S_{Eh_i}^G$ is (unlike Q) a genuine parabolic subgroup of G will now be used to show that their corresponding orbits are "negligible" .

Lemma (10.1.2) Outside of a finite set of values for q^{-s}, $Hom_G(\pi, ind_{Q^{h_i}}^G \tau^{h_i} \otimes ||^s) = 0$, for $i = 1, 2$.

Proof. This is once again an immediate Corollary of the general theory of induced representations for reductive groups as developed in [BZ] or [Casselman]. Because π is an irreducible admissible representation of G, there exists a parabolic subgroup $Q_1 = M_1 N_1$ of G and an irreducible cuspidal representation σ_1 of M_1 such that π is a subrepresentation of $Ind_{Q_1}^G \sigma_1 \otimes 1$; moreover, the "cuspidal data" (M_1, σ_1) is uniquely determined by π up to "conjugacy". Similarly, the irreducible representation $\tau^{h_i} \otimes ||^s$ of Q^{h_i} determines cuspidal data (P_2, σ_2^s) where P_2 is a parabolic subgroup of the Levi component of Q^{h_i} and σ_2^s is an irreducible cuspidal representation of the Levi component of P_2. Thus $dim \; Hom_G(\pi, Ind_{Q^{h_i}}^G \tau^{h_i} \otimes ||^s) \leq dim \; Hom(Ind_{Q_1}^G \sigma_1 \otimes 1, \; ind_{P_2}^G \sigma_2^s \otimes 1)$. By Theorem 2.8

and 2.9 of [BZ], the Jordan-Holder series of each of these latter induced representations is finite. Moreover, since π and τ are predetermined, there can be only finitely many possible values for $||^s$ (and hence q^{-s}) such that these Hom spaces are non-zero. Thus the Lemma is proved.

Remark. Lemma 10.1.2 is the local analogue of the fact (established in Section 3) that the orbits Eh_1 and Eh_2 are negligible so far as the global zeta-integral $I(s,\varphi,E)$ is concerned. Globally, the point is that φ cuspidal implies that the contribution of either of these orbits to $I(s,\varphi,E)$ vanishes. If we assume, in the present local context, that the representation π is itself actually cuspidal, then $Hom_G(\pi, ind_{Q^i}^G \tau^{h_i} \otimes ||^s)$ is identically 0 for all s (as opposed to almost all q^{-s}).

(10.2) Analysis of $ind_Q^G \tau \otimes ||^s$.

According to the last paragraph, our uniqueness theorem for G-invariant bilinear forms on $V_\pi \times V_{ind\ \tau\otimes||^s}$ will follow from the fact that

$$dim\ Hom_G(\pi, ind_Q^G \tau \otimes ||^s) = 1$$

except for finitely many values of q^{-s}.

Recall that $\tau \otimes ||^s$ is given to us initially as a representation of P. In order to analyze $ind_Q^G \tau \otimes ||^s$, we must therefore first study how $\tau \otimes ||^s$ decomposes upon restriction to $Q = P \cap G$. This, in turn, involves the Gelfand-Kazhdan (cf. [BZ]) theory of restricting representations of GL_n to P_n.

Suppose first that τ is a cuspidal representation of GL_n. In this case, $\tau \otimes ||^s|_{P_n}$ is irreducible and isomorphic to the standard representation $ind_{Z_n}^{P_n} \psi$. (This result - due to Gelfand-Kazhdan - was already described in terms of the theory of derivatives in section 7.2.) From this it follows that

$$ind_Q^G \tau \otimes ||^s = ind_{Z_n U^Q}^G \psi, \tag{10.2.1}$$

where $Q = GL_{n-1}U^{P^*} = P_n U^Q \supset Z_n U^Q$, and $Z_n U^Q$ is the maximal unipotent subgroup of G described in Remark 3.3.1. Thus

$$dim\ Hom(\pi, ind_Q^G \tau \otimes ||^s) = 1,$$

and our desired uniqueness theorem follows. Indeed (10.2.1) renders this last statement equivalent to the existence and uniqueness of Whittaker models for π.

For arbitrary τ, the situation is only slightly more complicated. As we shall see below, (10.2.1) will hold modulo certain "small" representations of H which will make negligible contribution to $Hom_G(\pi, ind_Q^G \tau \otimes \|^s)$. Recall that there exists some parabolic $P_{(n)}$ of GL_n, and some cuspidal representation τ' of the Levi component of $P_{(n)}$, such that τ is a subrepresentation of $ind_{P_{(n)}}^{GL_n} \tau'$. From the point of view of putting a bound on $dim\ Hom_G(\pi, ind_Q^G \tau \otimes \|^s)$, it therefore suffices to assume $\tau = ind_{P_{(n)}}^{GL_n} \tau'$.

Let $\sigma = \tau|_{P_n}$, where $\tau = ind_{P_{(n)}}^{GL_n} \tau'$. From the Basic Structure Theorem 7.2., we know that σ is "glued together" from representations of the form

$$ind_{R_{i-1}}^{P_n} \sigma^{(n-i+1)} \otimes \psi$$

for $i = 1, \cdots, n$, and $R_m \supset GL_m Z_{n-m}$. On the other hand, from Sections 4.4, 4.6 and 4.7 of [BZ], we also know that

(i) $\sigma^{(n)} = 1$; and

(ii) if $P_{(n)}$ has Levi component $GL_{n_1} \times \cdots \times GL_{n_r}$, and $\tau' = \rho_1 \otimes \cdots \otimes \rho_r$, with each ρ_i a cuspidal representation of GL_{n_i}, then there are only finitely many irreducible constituents of $\sigma^{(m)}$, $m = 0, 1, \cdots, n$, and each is such that its cuspidal data is a subset of $\rho_1, \cdots, \rho_r$.

From this it follows that $ind_Q^G \tau \otimes \|^s$ differs from $ind_{Z_n U}^G \psi$ by a collection of "small representations". Here "small representation of G" means one properly induced from a parabolic subgroup of G. More precisely, each such small representation is of the form

$$ind_{Q^G}^G (\sigma_m \otimes \|^s) \otimes \sigma'$$

where Q^G is a proper parabolic subgroup of G with Levi component isomorphic to $GL(m)SO(2(n-m))$, σ_m is a cuspidal representation of $GL(m)$, and σ' is a cuspidal representations of $SO(2(n-m))$. The crucial point is that there are only a finite number of possibilities for σ_m, and they are determined by τ.

On the other hand - again by Theorem 2.9 of [BZ], a given irreducible representation π of G can intertwine with such a "small" representation only if s is appropriately chosen (the choice being determined by π and τ). Thus we have proved

<u>**Theorem B.**</u> Outside of a finite set of values for q^{-s}, the space $Hom_G(\pi, ind_P^H \tau \otimes ||^s)$ is one-dimensional.

This establishes the functional equation for groups of type D_n.

§11. Groups of Type C_n

Our local zeta-integral now has the form

$$\varsigma(s, W, \Phi, f) = \int_{U^G \backslash G} W(g) F^{\Phi}(g) W_f^{\tau}(g, s) dg \tag{11.0.1}$$

where $W(g)$ belongs to the Whittaker model $\mathcal{W}(\pi, \psi)$, $F^{\Phi}(g) = r(g)\Phi(e_n)$ with Φ in $S(k^n)$, and $W_f^{\tau}(g, s)$ is such that the function $w(m) = W_f^{\tau}(mg, s)$ belongs to $\mathcal{W}(\tau, \psi^{-1})$ for each g. Recall that F^{Φ} and $W_f^{\tau}(g, s)$ are actually "genuine" functions on the metaplectic cover $Mp(k)$ of G, and $r(g)$ denotes the Weil representation.

The zeta-integral $\varsigma(s, W, \Phi, f)$ clearly defines a $Mp(k)$-invariant trilinear form on $\mathcal{W}(\pi, \psi) \times S(k^n) \times ind\ \tau \|\|^s$. Our task is to prove that this form is unique (at least for almost all q^{-s}). By the usual abstract nonsense relating multilinear forms with tensor products and contragredients, this problem is reduced to analyzing the dimension of the space

$$Hom\ (\pi \otimes r,\ (ind\ \tau\|\|^s)^{\sim})$$

where $\sim$ denotes the contragredient representation. Indeed, for each pair (v, w) in $V_\pi \times S(k^n)$, we can consider the bilinear Mp-map $\alpha : (v, w) \to (ind\ \tau\|\|^s)^{\sim}$ defined by $\alpha(v, w)(n) = \varsigma(s, v, w, n)$; thus the theory of tensor products leads us to the space just described above. Frobenius reciprocity and similar general arguments allow us to replace this last space by

$$Hom_P(\pi \otimes \tau\|\|^s, r^{\sim}).$$

Actually the contragredient of r is isomorphic to $r_{\psi^{-1}}$; similarly, $(ind\ \tau\|\|^s)^{\sim}$ is of the form $ind\ \tilde{\tau}\|\|^{-s}$. In any case, our first step is to analyze the restriction of r to P.

(11.1) Analysis of $r|_P$.

In Lemma 4.3.4 we defined a character ψ_U of $U^P = \left\{u_S = \begin{bmatrix} I_n & S \\ 0 & I_n \end{bmatrix} : {}^t s = s\right\}$ through the formula

$$\psi_U(u_S) = \psi(\rho_n(u_S)) = \psi(s_{nn})$$

if $S = (s_{ij})$. Note that ψ_U is a "degenerate character of U of *rank* 1 ", and the stabilizer of ψ_U in $M^P \approx GL_n$ (under the natural action of M^P on U^P given by conjugation) is P_n. Therefore ψ_U extends trivially across P_n to define a character ψ_U of $P_n U^P \subset P$. Henceforth, set

$$R = P_n \cup^P .$$

Proposition (11.1.1). The restriction of the Weil representation $r_{\psi^{-1}}$ to the parabolic subgroup $P = M \cup^P$ is isomorphic to the sum of $ind_R^P \psi_U$ and a one-dimensional representation which is the quotient in $S(k^n)$ of the hyperplane $\{\phi \in S(k^n) : \Phi(0) = 0\}$.

Remark. We have already noted in Section 4 that the restriction of r to P defines an ordinary representation of P. By "the sum of" we mean "glued from"; i.e. there is a subspace of $\tilde{r}|_P$ isomorphic to $ind\ \psi_U$, and the resulting quotient representation is one-dimensional. Correspondingly, we break up the proof of the Proposition into two steps.

Proof. (a) We shall first show that $ind\ \psi_U$ embeds as a subspace of $\tilde{r}|_P$. Recall that $ind\ \psi_U$ refers to "compactly supported induction", i.e., for each f in $ind\ \psi_U$, there exists a compact set K_f^P in P such that support $(f) \subset P_n \cup^P K_f$. Suppose now that $m_x = \begin{pmatrix} g_x & 0 \\ 0 & {}^t g_x^{-1} \end{pmatrix}$ in $M \subset P$ is such that $x \cdot g_x = e_n$. We claim then that the map

$$f \to \Phi_f(x) = |det\ g_x|^{-1/2} f(m_x)$$

intertwines $ind\ \psi_U$ with the subspace

$$S_0 = \{\Phi \in S(k^n) : \Phi(0) = 0\}$$

of $S(k^n)$.

Let us check that $\Phi_f(x)$ is well-defined. If m'_x is also such that $e_n = x \cdot g'_x$, then $g'_x = p_n g_x$ with p_n in P_n, since P_n is the stabilizer of e_n in $GL(n)$. Therefore

$$\begin{aligned} \Phi'_f(x) &= |det\ g'_x|^{-1/2} f(m'_x) \\ &= |det\ g_x|^{-1/2} |det\ p_n|^{-1/2} |det\ p_n|^{1/2} f(m_x) \\ &= \Phi_f(x), \end{aligned}$$

as required.

It is clear that $\Phi_f(x)$ is locally constant (since f is) and that Φ_f vanishes for x large or small (since f is compactly supported modulo $P_n U^P$) ; it remains to check that $f \to \phi_f(x)$ is an embedding of P-modules.

First suppose $\Phi_f \equiv 0$. Since all $x \neq 0$ in k^n are in the $GL(n)$-orbit of e_n (indeed this orbit is isomorphic to $P_n \setminus GL(n)$), this means that $f \equiv 0$. Thus $f \to \phi_f$ is $1-1$. To verify the required intertwining, pick $m = \begin{pmatrix} g_m & 0 \\ 0 & {}^t g_m{}^{-1} \end{pmatrix}$ arbitrary in $M \approx GL(n)$. Then $g_m^{-1} g_x$ takes $x g_m$ to e_n, and therefore

$$\begin{aligned} \Phi_{m \cdot f}(x) &= |det\ g_x|^{-1/2}(m \cdot f)(\cdot m_x) = |det g_x|^{-1/2} f(m^{-1} m_x) \\ &= |det\ g_x|^{-1/2} |det g_m^{-1} g_x|^{1/2} \Phi_f(x \cdot g_m) \\ &= |det\ g^{-1}|^{1/2} \Phi_f(x \cdot g_m) \end{aligned}$$

I.e., on M at least, $f \to \Phi_f$ intertwines $ind\ \psi_U$ with the Weil representation $\tilde{r}$. Similarly, suppose $u = \begin{bmatrix} I_n & S \\ 0 & I_n \end{bmatrix}$ is arbitrary in U^P. Then

$$\begin{aligned} \Phi_{uf}(x) &= |det\ g_x|^{-1/2} f(u^{-1} m_x) = |det\ g_x|^{-1/2} f(m_x(m_x^{-1} u^{-1} m_x)) \\ &= |det\ g_x|^{-1/2} \psi_U(-g_x^{-1} S g_x) f(m_x) \\ &= \psi(-x S^t x) \Phi_f(x) = \tilde{r}(u) \Phi_f(x) \end{aligned}$$

since $x = e_n \cdot g_x^{-1}$, and $p_n(u) = e_n S^t e_n$.

(b) To complete the proof of Proposition 11.1.1, we need to show that the image of $f \to \Phi_f$ is the hyperplane S_0 of $S(k^n)$ (consisting of all functions which vanish at 0). We do this by constructing an inverse map $\Phi \to f_\Phi$ from S_0 to $ind\ \psi_U$ as follows. If $p = m_p u_p$ in $P = MU^P$, define f_Φ on P by

$$f_\Phi(p) = |det\ g_p|^{1/2} \psi_U(u_p) \Phi(e_n g_p).$$

Here, as before, $m_p = \begin{pmatrix} g_p & 0 \\ 0 & {}^t g_p{}^{-1} \end{pmatrix}$, $g_p \in GL(n)$. The claim is that f_Φ belongs to $ind\ \psi_U$, and the map $\Phi \to f_\Phi$ is a P-module isomorphism with inverse $f \to \Phi_f$. The verification of these assertions is completely straightforward. For example, to see that f_Φ transforms properly on the right with respect $m = \begin{pmatrix} g & 0 \\ 0 & {}^t g^{-1} \end{pmatrix}$, $g \in P_n$, we compute

$$\begin{aligned} f_\Phi(pm) &= f_\phi(m_p u_p m) = f_\Phi(m_p m(m^{-1} u_p m)) \\ &= |det\ g_p g|^{1/2} \psi_U(m^{-1} u_p m) \Phi(e_n (g_p g)^{-1}) \\ &= |det\ g|^{1/2} |det\ g_p|^{1/2} \psi_U(u_p) \Phi(e_n g_p^{-1}) \\ &= |det\ g|^{1/2} f_\Phi(p) \end{aligned}$$

since P_n stabilizes e_n and ψ_U .

Proposition 11.1.2. Outside of a finite number of values for q^s,

$$Hom_P(\pi \otimes \tau||^s, \tilde{r})$$

$$= Hom_R(\pi \otimes \tau||^s, \psi_U).$$

Proof. Recall that R denotes the subgroup $P_n U^P$ of P. In the last Proposition we proved that $\tilde{r}|P$ is glued together from its submodule $ind_R^P \psi_U$ and a one-dimensional quotient χ_1. Clearly

$$Hom_R(\pi \otimes \tau||^s, \chi_1) = \{0\}$$

unless $||^s$ (and hence q^s) is properly chosen (depending on π, τ, and χ_1, all of which are fixed). On the other hand, by Frobenius reciprocity,

$$Hom_P(\pi \otimes \tau||^s),\ ind_R^P \psi_U) \approx Hom_R(\pi \otimes \tau||^s, \psi_U).$$

Therefore, outside of a finite set of values for q^{-s}, $Hom_P(\pi \otimes \tau||^s, \tilde{r}) \approx Hom_R(\pi \otimes \tau||^s, \psi_U)$, as claimed.

(11.2) Proof of the Functional Equation

We have explained how the functional equation for $\varsigma(s, W, \Phi, f)$ reduces to the assertion that the dimension of $Hom_P(\pi \otimes \tau||^s, \tilde{r})$ is 1 (except for possibly finitely many values of q^{-s}). By the paragraph above, this is equivalent to a similar assertion for the space $Hom_R(\pi \otimes \tau||^s, \psi_U)$. Equivalently, it remains to prove:

Proposition (11.2) Except for finitely many values of q^{-s},

$$dim\ Hom_R(\pi, \tau||^s \psi_U) = 1.$$

Proof. Suppose first that τ (and hence $\tau||^s$) is a cuspidal representation of $GL_n(k)$. Then (cf. Section 7.2) $\tau||^s$ restricted to P_n is of the form $ind_{Z_n}^{P_n}\theta$ (with θ the standard non-degenerate character of Z_n defined by $\theta((z_{ij})) = \psi(\sum_{i=1}^{n-1} z_{i,i+1}))$. Thus $\tau||^s\psi$, regarded as a representation of P and restricted to $R = P_nU^P$, is of the form $ind_{Z_nU^P}^{R}\theta\psi_U$.

By Frobenius reciprocity,

$$Hom_R(\pi, \tau||^s\psi_U) \approx Hom_{Z_nU^P}(\pi, \theta\psi_U).$$

Since $\theta\psi_U$ defines a non-degenerate character of the maximal unipotent subgroup $U^G = Z_nU^P$ of G, the required uniqueness result is merely a restatement of the uniqueness theorem for standard Whittaker models for π.

In the general case, we want to argue similarly, exploiting the fact that $\tau|_{P_n}$ differs from *ind* θ only by "small representations" which generically contribute nothing to $Hom_R(\pi, \tau||^s\psi_U)$. Recall first that R is the semidirect product of P_n with U^P; thus we may apply the Jacquet functor r_{U^P,ψ_U} to π in order to replace $Hom_R(\pi, \tau||^s\psi_U)$ by $Hom_{P_n}(r_{U^P,\psi}(\pi), \tau||^s)$. (cf. §3.2 of [Casselman]). Because U^P is the unipotent radical of the parabolic subgroup $P \approx GL(n)\cup^P$ in G, the Jordan-Holder series of $r_{\cup,\psi}(\pi)$ is finite (as a $GL(n)$-module). Moreover, the restriction of each irreducible constituent of $r_{\cup,\psi}(\pi)$ to P_n has constituents of the form $ind_{R_m}^{P_n}\sigma_*^m \otimes \psi$, where $0 \leq m \leq n-1$, and σ_*^m is a representation of $GL(m)$ (the $(n-m)$-th derivative of the restriction of the appropriate constituent of $r_{\cup,\psi}(\pi)$). (Cf. §(7.2).) Similarly, as in Section 10.2, we may assume that the restriction of $\tau||^s$ to P_n is glued together from representations of the form $ind_{R_m}^{P_n}(\sigma^{(n-m)} \otimes ||^s) \otimes \psi$, where $\sigma^{(n-m)}$ is the $(n-m)$-th derivative of $\tau = (ind^{GL(n)}\tau')|_{P_n})$, τ is a cuspidal representation of some parabolic subgroup $GL(n)$, and there are only finitely many irreducible constituents of any $\sigma^{(m)}, m = 0, \cdots, n-1$, each one of which has cuspidal data contained in the cuspidal data of τ'. Thus it is clear that - outside of a finite set of values for q^{-s}— the derivatives of order $m \geq 1$ make no contribution to $Hom_{P_n}(r_{\cup,\psi_U}(\pi), \tau||^s)$. By the theory of Whittaker models, the proof of Proposition 11.2 is therefore complete.

§12. Rationality of the Zeta-functions

In the previous five Sections, we have established a functional equation for the various local zeta-integrals $\varsigma(s, W, f)$. For example, for groups of type B_n, this equation reads

$$\varsigma(1-s, W, M(s)f) = \gamma(s, \pi \times \tau, \psi)\varsigma(s, W, f),$$

where $M(s)f$ belongs to $ind\ \tilde{\tau} \otimes ||^{-s'}$. As already pointed out in §7.1, what we actually prove is that the space of H-invariant bilinear forms on $V_\pi \times V_{Ind\ \tau||^s}$ is one dimensional (outside of a finite set of values for q^{-s}) and spanned by $\varsigma(s, W, f)$ for $Re(s)$ sufficiently large.

Since we are assuming that these zeta-integrals, initially defined only in some right half-plane, are actually meromorphic in the whole s-plane, our desired functional equation results immediately from this uniqueness result.

It is the purpose of the present Section to finally establish the meromorphy of $\varsigma(s, W, f)$; in fact, we shall show that these zeta-integrals define rational functions of q^{-s}.

(12.1) Traditionally, rationality results for zeta-integrals involving Whittaker functions are proved by appealing to explicit asymptotic expansions for these functions. For example, in the pioneering work of [Jacquet] on Rankin-Selberg convolutions for $GL(2) \times GL(2)$, the local zeta-integrals are of the form

$$\varsigma(s, W_1, W_2, f) = \int_{UZ\backslash GL_2(k)} W_1(g)W_2(\eta g)f(g,s)dg$$

where $\eta = \begin{bmatrix} -1 & 0 \\ 0 & 1 \end{bmatrix}$, $U = \left\{\begin{bmatrix} 1 & b \\ 0 & 1 \end{bmatrix}\right\}$, $Z = \left\{\begin{bmatrix} c & 0 \\ 0 & c \end{bmatrix}\right\}$, W_i belongs to the Whittaker model of an irreducible representation π_i of $GL_2(k)$, and $f(g, s)$ belongs to an induced representation of $GL_2(k)$ such that $f\left[\left[\begin{pmatrix} a_i & b \\ 0 & a_2 \end{pmatrix}\right] g, s\right] = |a_1|^s |a_2|^{-s} f(g, s)$. Since all the representations of GL_2 involved here are admissible, the functions in the integrand are invariant by an open compact subgroup of $K = GL_2(0_k)$, and hence by Iwasawa's decomposition, $\varsigma(s, W_1, W_2, f)$ is a finite linear combination of integrals of the form

$$\int_{k^x} W_1^*\begin{pmatrix} a & 0 \\ 0 & 1 \end{pmatrix} W_2^*\begin{pmatrix} -a & 0 \\ 0 & 1 \end{pmatrix} |a|^{s-1} d^x a$$

where each $W_i^*\left(\begin{smallmatrix} a & 0 \\ 0 & 1 \end{smallmatrix}\right) = \xi_i(a)$ belongs to the Kirillov model of π_i. But now the result on asymptotic expansions for Whittaker models alluded to above implies precisely that any such function $\xi(a)$ must be of the form

$$\sum_{i=1}^{N} \Phi_i(a)\lambda_i(a) \tag{12.1.1}$$

with Φ_i in $S(k)$ and λ_i a "finite" function on k^x, i.e., a finite linear combination of functions of the form $\mu(a)v(a)^m$ with μ a character of k^x, v the discrete valuation on k, and m a non-negative integer. Since the integral of such functions against $|a|^s$ clearly defines a rational function of q^{-s}, the proof of rationality for $\varsigma(s, W_1, W_2, f)$ is complete.

In the present general context, we could try to proceed similarly by appealing to an analogue of (12.1.1); cf. [JPSS1], 2.7 for the case of GL_n. Note that (12.1.1) is an "asymptotic expansion" because it describes the behaviour of $W\left(\begin{smallmatrix} a & 0 \\ 0 & 1 \end{smallmatrix}\right)$ as a approaches 0; in fact, the claim is that as a approaches 0, $W\left(\begin{smallmatrix} a & 0 \\ 0 & 1 \end{smallmatrix}\right)$ behaves like a finite sum of characters of k^x. In general, the proof of rationality along these lines will be more complicated since there will be more ways (corresponding to the n different simple roots) of approaching 0. Therefore, as promised, we want to appeal to a simpler and more elegant approach due to J. Bernstein.

(12.2) Bernstein's theory is remarkably general as well as broad in its range of applicability ([Bernstein]). The theory seems to have been devised primarily for applications to the analytic continuation of intertwining operators and Eisenstein series (cf. the remarks on p.67 of [Kaz Pat]). We proceed now to describe the set-up of the theory.

Let Y be a vector space over a field K, and $Y^* = Hom_K(Y, K)$ its linear dual. By a system of equations Ξ in Y^* we understand a collection of pairs $\{(y_\nu, \lambda_\nu)\}$ indexed by some set $R = \{\nu\}$, where $y_\nu \in Y$, and $\lambda_\nu \in K$. A solution of Ξ is then a vector ℓ in Y^* such that $\ell(y_\nu) = \lambda_\nu$ for all ν in R. Now let D be an algebraic variety defined over K, with ring of regular functions $K[D]$ over K. A function $f : D \to Y$ will be called regular (or polynomials) on D if it is given by an element of $Y \otimes_K K[D]$, i.e., if it is a finite sum $\sum y_i \varphi_i$ with φ_i in $K[D]$ and y_i in Y.

Finally, suppose $\{\Xi_d\}$ is a family of systems of equations in Y^*parametrized by d in D. We say $\{\Xi_d\}$ is polynomial in d if all the systems Ξ_d are indexed by the same set$R = \{\nu\}$, and for each ν, $x_\nu(d)$ (resp. $\lambda_\nu(d)$) is regular on D, i.e., belongs to $Y \otimes K[D]$ (resp. $K[D]$. Before stating the fundamental theorem, we introduce some more notation. Let L denote the field of fractions of $K[D]$, $Y_L = Y \otimes_K L$, and $Y_L^* = Hom_L(Y_L, L)$, the linear dual of Y_L over L. Note that we regard the system of equations $\Xi = \{(Y_\nu, \lambda_\nu)\}$ as a system in Y_L^* by viewing y_ν as an element of Y_L, and λ_ν as an element of L.

Bernstein's Theorem. Suppose we are given a polynomial family $\{\Xi_d\}$with $K = \mathbb{C}$, D an irreducible variety over $\mathbb{C}$, and Y of countable dimension over $\mathbb{C}$. Suppose also that there exists a subset $\Omega \subset D$ nonempty and open (in the usual Hausdorf topology) such that for each d in Ω, the system Ξ_d has a unique solution ℓ_d. Then the system Ξ has a unique solution ℓ in Y_L^*, and on some subset $D' \subset D$, whose complement is a countable union of hypersurfaces, $\ell(d)$ is the unique solution ℓ_d of Ξ_d.

(12.3) We shall use Bernstein's theorem to prove the following :

Proposition 12.3. For any W and f ,

$$\varsigma(s, W, f) = \int_{\cup\backslash H} W(h) f(h, s)\, dh$$

defines a rational function of q^{-s}. In particular, $\varsigma(s, W, M(s)f)$ is well-defined for all s in $\mathbb{C}$, and again a rational function of s.

We start by reformulating our result that $Hom_H(\pi, Ind_P^H\, \tau \otimes ||^s)$ is one-dimensional. Let Y denote the tensor product space $V_\pi \otimes V_{Ind_P^H\, \tau}$. Note that for any s, the space of the induced representation $Ind\, \tau \otimes ||^s$ is naturally isomorphic to $V_{Ind\, \tau}$ (the isomorphism arising from the restriction of functions in the induced space to the maximal compact subgroup K of H). However, the action of H on $V_{Ind\, \tau}$ does depend on s. More precisely, if $kh_0 = pk'$, with h_0 in H, then h_0 acts on φ in $V_{Ind\ \tau}$ through the formula

$$\Pi_s(h_0) : \varphi(k) \to \tau(p)\varphi(k')|det\ m|^{s*}$$

if $p = mu$. (Here Π_s denotes the representation $Ind\ \tau \otimes ||^s$; the translation $s \to s^* = s + \frac{n-1}{2}$ arises from the modulus function on P.) On the space $Y = V_\pi \otimes V_{Ind\ \tau}$, H acts through the formula

$$\Pi_s(h_0)(v \otimes \varphi) = \pi(h_0)v \otimes \Pi_s(h_0)h.$$

In order to reformulate our result on the uniqueness of H-invariant bilinear forms on $V_\pi \times V_{Ind\ \tau\otimes||^{s'}}$ in terms of the tensor product space $Y = V_\pi \otimes V_{Ind\ \tau}$, let us remember that bilinear forms on $V_\pi \times V_{Ind}$ are the same things as linear forms on $V_\pi \otimes V_{ind\ \tau}$. Let us suppose also that W_0 and f_0 are functions in V_π and $V_{ind\ \tau\otimes||^{s'}}$ with the property that $\varsigma(s, W_0, f_0) = 1$ (the existence of such functions is of independent interest, and proved in Proposition 12.4 below); then we have

Proposition (12.3.1) Outside of a finite set of values for q^{-s}, there is a unique (non-trivial) linear functional $\ell_s : Y \to \mathbb{C}$ such that

$$\ell_s(\Pi_s(h)y - y) = 0 \ \underline{\text{and}} \ \ell_s(W_0 \otimes f_0) = 1.$$

In particular, for $Re(s)$ sufficiently large, this functional is given by the integral $\varsigma(s, W, f)$, if $y = W \otimes f$; moreover, this solution depends analytically on s in this domain.

In order to bring into play Bernstein's theory, we need to view (12.3.1) as a <u>family</u> of systems of equations in $Y^* = Hom_{\mathbb{C}}(Y, \mathbb{C})$. So let D denote the multiplicative group of $\mathbb{C}$ regarded as an irreducible algebraic variety over $\mathbb{C}$. We parametrize each non-zero complex number z in D by $z = q^{-s}$, with $-\frac{\pi}{\ell nq} \leq Im(s) < \frac{\pi}{\ell nq}$, so that $\mathbb{C}[D] = \mathbb{C}[z, z^{-1}] = \mathbb{C}[q^{-s}, q^s]$. We also let R_Y (resp. R_H) index a countable basis for the space Y (resp. the group H). Then for each $z = q^{-s}$ in D, consider the collection of pairs $\Xi_s = \{\Pi_s(h_\nu)y_{\nu'} - y_{\nu'}, 0\}$ indexed by (ν, ν') in $R = R_Y \times R_H$. By the Proposition above, the system Ξ_s has a unique solution ℓ_s in Y^*, provided q^{-s} avoids a finite set of values in $\mathbb{C}$. Note, moreover, that the family of systems of equations $\{\Xi_s\}$ is obviously polynomial in z, since $\Pi_s(h_\nu)y_{\nu'} - y_{\nu'}$ clearly belongs to $Y \otimes \mathbb{C}[q^{-s}, q^s]$. Therefore, all the hypotheses of Bernstein's theorem are satisfied , and its conclusion implies the existence of a linear functional

$$\ell : Y \otimes L \to L,$$

with L the field of fractions of $\mathbb{C}[D]$, and $\ell(y \otimes q^{-s}) = \ell_s(y)$ for all y in Y, and all but finitely many values of q^{-s} .

Now for $Re(s)$ sufficiently large,

$$\varsigma(s, W, f) = \int_{U \backslash H} W(h) f(h, s) dh$$

converges and defines a linear functional $\varsigma(s)$ on $V_\pi \otimes V_{Ind\ \tau} = Y$ such that $\varsigma(s)(W_0 \otimes f_0) = 1$ and $\varsigma(s)(\Pi_s(h)y) = \varsigma(s)(y)$, i.e., $\varsigma(s) = \ell_s$. Therefore, by the conclusion above, $\varsigma(s, W, f) \in \mathbb{C}[q^{-s}, q^s]$ as a function of s, and we have finally proved

Proposition 12.3. The zeta integrals $\varsigma(s, W, f)$ are rational functions of q^{-s}.

(12.4) It remains to prove the following :

Proposition 12.4. There exists a W_0 in $\mathcal{W}(\pi, \psi)$, and an f_0 in $ind\ \tau \otimes \|\,\|^s$, such that

$$\varsigma(W_0, f_0, s) = \int_{U^H \backslash H} W_0(h) f_0(h, s) dh \equiv 1.$$

We note that this result will ultimately play a role in defining the local factor $L(s, \pi \times \tau)$ as a "g.c.d." of the integrals $\varsigma(W, f, s)$, even when π and τ are ramified. We hope to return to this question in future works.

Proof of Proposition 12.4. Recall that any $f^\tau(h, s)$ in $ind\ \tau \otimes \|\,\|^s$ is such that for each fixed h in H, the function $w(m) = f^\tau(mh, s)$ belongs to $\mathcal{W}(\tau \otimes \|\,\|^{s*}, \psi^{-1})$. We may assume $f_0(h, s)$ in $ind\ \tau \otimes \|\,\|^{s'}$ chosen so as to have support in PK_0, where K_0 is a compact open subgroup of H and K_0 is small enough so that both $f_0(h, s)$ and $W(h)$ are right K_0 invariant. Using the decomposition $H = PK_H$, we may rewrite $\varsigma(W, f, s)$ as

$$\int_{U^H \backslash P} W_\pi(p) f_0(h, s) |det\ p|^* dp$$

$$= \int_{Z_n \backslash GL_n} W_\pi(m) W_\tau^0(m) |det\ m|^{s*} dm$$

where s^* is an appropriate translate of s (determined by the modulus function on $P = MU^P$) and $W_\tau^0(m)$ belongs to the Whittaker model $\mathcal{W}(\tau, \psi^{-1})$. It remains to analyze the space of functions $\{W|_p : W \in \mathcal{W}(\pi, \psi)\}$.

Recall that $U^P \setminus Q \approx P_{n+1}$ via the projection $\alpha : Q \to P_{n+1}$ described in Proposition 2.2. According to the Lemma below, we may choose $W = W_0$ so that $W_0|_Q$ has arbitrary small compact support modulo U^P. In particular, we may choose W_0 to be the "delta-function at the identity", so that $W_0(m) W_\tau^0(m) = 0$ unless m belongs to a tiny neighbourhood of e in $GL_n(0_p)$, and on this neighbourhood equals 1. Since $|det\ m|^{s*} \equiv 1$ on this neighbourhood, $\varsigma(W_0, f_O, s)$ is indeed a constant (independent of s), and the Proposition is proved.

<u>**Lemma (12.5)**</u> For each $W \in \mathcal{W}(\pi, \psi)$, define a function W^* on P_{n+1} through the formula

$$W^*(p) = W(\alpha^{-1}(p)).$$

Then the map

$$W \longrightarrow W^*$$

is a homomorphism of P_{n+1} modules whose image contains $ind_{Z_{n+1}}^{P_{n+1}} \psi$.

<u>**Proof.**</u> Since the kernel of α is U^P, and ψ (and hence W) is invariant by U^P, this function is well-defined. The action of P_{n+1} on W^* is given by right translation, and it is clear that $W^*(zp) = \psi(z) W^*(p)$ for z in Z_{n+1}. Indeed, $W^*(zp) = W(\alpha^{-1}(z)\alpha^{-1}(p)) = \psi_{U^G}(\alpha^{-1}(z)) W^*(p)$, where $\alpha^{-1}(Z_{n+1}) = U^G$, and ψ_{U^G} is the non-degenerate character of U^G defined through the formula $\psi_{U^G}(zu) = \psi(\sum_{i=1}^n z_{i,i+1})$, z in Z_{n+1}, u in U^P.

Let $Ind_{Z_{n+1}}^{P_{n+1}} \psi$ denote the space of <u>all</u> locally constant functions on P_{n+1} which transform under the left action of Z_{n+1} according to the character ψ.

The image space

$$\{W^* : W \ \ in \ \ \mathcal{W}(\pi, \psi)\}$$

is then a (P_{n+1}) invariant subspace of $Ind_{Z_{n+1}}^{P_{n+1}}\psi$. Thus, according to Theorem F of [Gel Kaz], this subspace must contain $ind_{Z_{n+1}}^{P_{n+1}}\psi$ (the irreducible subspace of $Ind\ \psi$ consisting of functions compactly supported module Z_{n+1}). This completes the proof of Lemma (12.5), and hence the rationality of $\varsigma(s, W, f)$.

§13. Normalization of the Eisenstein Series

Recall that (at least for groups of type B_n) our local zeta-integral is of the form

$$\varsigma(s, W, f) = \int_{U\backslash H} W(h) W^f(h, s)\,dh,$$

where $W(h)$ belongs to the Whittaker model for π on G, and f belongs to the representation $ind_P^H \tau \otimes \|^{s'}$, with τ a non-degenerate representation of $GL(n,k)$ and $s' = s - 1/2$; the corresponding functional equation takes the form

$$\varsigma(1-s, W, M(s)f) = \gamma(s, \pi \times \tau, \psi)\varsigma(s, W, f)$$

where $M(s)$ is the intertwining operator between $ind\, \tau\|^{s'}$ and the representation $ind\, \tilde{\tau}\|^{-s'}$.

The following general features of this setup make possible an analysis of the automorphic L-functions attached to $G \times GL(n)$:

(A) For "unramified data" W and f, the zeta-integral $\varsigma(s, W, f)$ coincides with the local Langlands factor $L(s, \pi \times \tau)$ attached to the (unramified) representation $\pi \times \tau$ on $G \times GL(n)$;

(B) The local functional equation is compatible with the functional equation for the global zeta-integral $I(s, \varphi, E_f)$ (recall that E_f denotes a parabolically induced Eisenstein series, and $I(s, \varphi, E_f)$ factors as a product of the local zeta-integrals $\varsigma(s, W, f)$); moreover, $I(s, \varphi, E_f)$ should extend to a meromorphic function in all of $\mathbb{C}$, with only finitely many poles;

(C) In general, the g.c.d. of the integrals $\varsigma(s, W, f)$ should define an Euler factor $L(s, \pi \times \tau)$ for arbitrary (non-degenerate) π and τ, hence also an ε-factor $\varepsilon(s, \psi, \pi \times \tau)$; moreover, these factors should be "factorizable" when both π and τ are induced.

Unfortunately, Part (C) of the program just outlined has not yet been completed. Our purpose in this Section is to normalize our Eisenstein series so that these assertions have at least a chance of being correct as stated. The computation of the local unramified integrals is reserved for the Appendix to this paper.

(13.1) Returning once again to the global situation, consider the Eisenstein series

$$E_f(h,s) = \sum_{P_k \backslash H_k} f(\gamma h, s).$$

Langlands' general theory of Eisenstein series ([Langlands 2,3]) applies to this special situation, and in fact simplifies considerably in this case. Roughly speaking, the theory describes the meromorphic properties of $E_f(h,s)$ in terms of those of the intertwining operator $M(s)$ taking f in $Ind\ \tau \otimes ||^{s'}$ to $M(s)f$ in $Ind\ \tilde{\tau} \otimes ||^{-s'}$.

To be more precise, recall that $P = MU^P$ is a maximal parabolic subgroup of H, $M \approx GL(n)$, τ is an automorphic cuspidal representation of $GL(n, \mathbb{A})$, and f in $ind_{P_{\mathbb{A}}}^{H_{\mathbb{A}}} \tau \otimes ||^{s'}$ is such that - for each h in $H_{\mathbb{A}}$ - the function $f(mh,s)$ belongs to the space of cusp forms realizing $\tau \otimes ||^{s' + \frac{n-1}{2}}$. Let w denote the element of the Weyl group of H which takes P to the "opposite" parabolic subgroup P^-; in particular, w conjugates $\begin{bmatrix} g & 0 \\ 0 & {}^t g^{-1} \end{bmatrix}$ into $\begin{bmatrix} {}^t g^{-1} & 0 \\ 0 & g \end{bmatrix}$, and the intertwining operator $M(w,s)$ corresponding to it is just the operator $M(s)$ alluded to above. For a more comprehensive discussion of these intertwining operators, see [Arthur]; his exposition of the results of [Langlands 2,3] also provides the proper background for the next few paragraphs.

Now suppose f is of the form $\otimes f_v$, where each f_v lies in $ind_{H_v}^{P_v} \tau_v \otimes ||_v^{s'}$, and for almost every finite place v, f_v is the (unique normalized) spherical function in $ind\ \tau_v ||^{s'}$. Then we may write

$$M(s)f = \otimes(M_v(s)f_v)$$

where each $M_v(s)$ is the local intertwining operator defined by the same kind of (convergent) integral globally defining$M(s)$. At the unramified places of $\tau = \otimes \tau_v$, $M_v(s)$ must map the (normalized) spherical vector f_v° to a multiple $m_v(w, s, \tau_v)$ of the unique zonal spherical vector in $ind\ \tilde{\tau}_v \otimes \|\|^{-s'}$. Thus we introduce the scalar valued function

$$m_S(s) = \prod_{v \notin S} m_v(w, s, \tau_v),$$

where S denotes the (infinite and) ramified primes for σ.

According to Langlands' theory, $E_f(g,s)$ continues to a meromorphic function in the whole s-plane satisfying a functional equation relating $E_f(g,s)$ to $E_{M(s)f}(g, 1-s)$. (Recall that the substitution $s \to 1-s$ corresponds to the change of variable $s' \to -s'$, since $s' = s - 1/2$.) The fact that $M(s)$ enjoys the same meromorphic continuation as $E_f(g,s)$ means that the analytic properties of $E_f(g,s)$ are bound together with those of $m_S(s)$. It remains to identify $m_S(s)$ with an appropriate Langlands L-function which is attached to the Levi component of P and described in terms of the theory of L-groups.

Recall that the L-group of $H = SO(2n)$ is the complex Lie group ${}^LH = SO_{2n}(\mathbb{C})$, and there is a one-to-one correspondence between parabolic subgroups of H and ${}^LH$. If ${}^LP$ denotes the parabolic subgroup of ${}^LH$ corresponding to P in H, then ${}^LP$ is a maximal parabolic subgroup of ${}^LH$ with Levi decomposition ${}^LMU$. Here ${}^LM$ is the L-group of M, namely $GL_n(\mathbb{C})$, and the unipotent radical U of ${}^LP$ is isomorphic to the subgroup

$$\left\{ \begin{bmatrix} I_n & X \\ 0 & I_n \end{bmatrix} : {}^tX = -X \right\}$$

of $SO_{2n}(\mathbb{C})$. Let r_M denote the natural representation of ${}^LM$ given by its adjoint action on $U : X \to g \cdot X \cdot g^t$. Equivalently, in classical language, r_M is the <u>antisymmetric</u> square representation Λ^2- the restriction of $GL_n(\mathbb{C})$ to the antisymmetric tensors in $\mathbb{C}^n \otimes \mathbb{C}^n$.

In terms of this L-group data, $m_S(s)$ may be written as

$$\text{(13.1.1)} \qquad \frac{L_S(2s', \tau, r_M)}{L_S(2s'+1, \tau, r_M)} = \frac{L_S(2s-1, \tau, r_M)}{L_S(2s, \tau, r_M)}$$

where $L_S(s, \tau, r_M)$ is the product of the local Langlands factors

$$L(s, \tau_v, r_M) = \{det\ [I - r_M(t_{\tau_v}) q^{-s}]\}^{-1}$$

taken over all the unramified places v not in S. Indeed, this can be read off from the formulas in §5 of [Arthur], keeping in mind that P is maximal parabolic and there is only one root subgroup of H inside U^P (equivalently, only one simple root of H whose restriction to the centre of M is non-trivial).

(13.2) For groups of type D_n, similar arguments show that $m_S(s)$ is of the form (13.1.1) with $r_M = Sym^2$, the symmetric square of the standard representation of $GL_n(\mathbb{C})$. For groups of type C_n, a computation of $M(s)$ is required from scratch, since $E_f(g,s)$ is now defined on the metaplectic cover of $Sp(n)$ rather than on the algebraic group $Sp(n)$ itself. In this case (13.1.1) holds with r_M again isomorphic to Sym^2 (rather than the representation (Standard) $\oplus\Lambda^2$ attached to the analogous Eisenstein series on $Sp(n)$ itself). The idea is to apply the method of Gindikin and Karpelevic exactly as in [Langlands 2], thereby reducing matters to the case of a simple reflection; in this case, the computation involves only (the metaplectic cover of) $Sp(1)$, i.e. Section 7 of [GeJa].

(13.3) In order to derive more detailed information on the poles of $E_f(g,s)$, one needs either to refine the methods of [Langlands 2]or the subsequent theory developed by Shahidi in his work on "local coefficients"; cf. [Shahidi]. Because of the relation (13.1.1), and its analogues for groups of type C_n and D_n, it seems reasonable to define a normalized Eisenstein series through the formula

$$E_f^N(g,s) = L_S(2s,\tau,r_M)E_f(g,s).$$

At least for properly chosen f, this normalized Eisenstein series should then have no more poles than the global Langlands L-functions appearing in the numerator of (13.1.1). In particular, the desired result (B) on the finiteness of the number of poles for $L_S(s,\pi\times\tau)$ should then follow from the verification of the analogous Langlands conjecture for the lower dimensional L-functions $L(s,\tau,\Lambda^2)$ or $L(s,\tau,Sym^2)$. These L-functions, in turn,, seem to be within the range of attack of our present understanding of L-functions attached to $GL(n)$. More precisely, we understand that F. Shahidi can prove that these L-functions have only finitely many poles; cf. [Shahidi 2]. Moreover, at least for n even and $r_M = \Lambda^2$, Jacquet and Shalika actually know how to locate these poles, using an explicit integral expression for $L(s,\tau,\Lambda^2)$. (For all these assertions, we are still assuming τ is non-degenerate.)

In closing, we note the relation between our approach to Rankin-Selberg convolutions and previous works on the subject. The prototype for all previous (representation theoretic) generalizations of the classical method of Rankin and Selberg is Jacquet's treatment of the case $GL(2) \times GL(2)$. In [Jacquet], the Eisenstein series are defined so that their normalization is already built into the choice of functions $f(h, s)$. In particular, for almost all v, the restriction of f_v to K_v is already the product of (a function like) $L(2s, \tau, r_M)$ times the constant function. This means, on the one hand, that the global Eisenstein series are already normalized so as to have only a finite number of poles. On the other hand, it also means that the (local) definition of $L(s, \pi \times \tau)$ as the g.c.d. of the local zeta-integrals will be the correct one, not only at the unramified places, but also at the bad primes. In their recent work on Rankin triple products for $GL(2)$, Piatetski-Shapiro and Rallis have sufficiently modified Jacquet's method of normalization of the Eisenstein series (and intertwining operators). Thus they are able to define local factors at an arbitrary prime, and locate the poles of their new global L-functions. It seems possible to carry out a similar program in the general context of the present paper, and we hope to do so in the near future.

Appendix

by

S. Gelbart, I. Piatetski-Shapiro and S. Rallis

The purpose of this appendix is to compute the "unramified" local zeta-integrals introduced in the preceding paper ([GePS4]); for definitions and notation, we refer the reader to Chapter I of that paper. Once again, we give details only for the case of groups of type B_n. Thus $G = SO(V) \supset SO(V') = H$, and π (resp. τ) is an unramified irreducible representation of G (resp. $GL_n(k)$) with Whittaker model $\mathcal{W}(\pi,\psi)$, (resp. $\mathcal{W}(\tau,\psi^{-1})$). Here ψ is a non-trivial character of the local field k with conductor O_k, the ring of integers of k, and our goal is to prove the following:

Proposition A.1. Suppose W (resp. W^f) is the unique function in $\mathcal{W}(\pi,\psi)$ (resp. $ind\ \tau \oplus \|^s)$ which is right invariant by the standard maximal compact subgroup K of G (resp. H) and normalized so that $W(e) = 1$. Then

$$(A.1.1) \qquad L(2s,\tau,\Lambda^2)\varsigma(s,W,f) = \frac{1}{det(I - r(t_\pi \otimes t_\tau)q^{-s})}.$$

Here t_π (resp. t_τ) denotes a representative of the semi-simple conjugacy class in $Sp_{2n}(\mathbb{C}) = {}^LG^\circ$ (resp. $GL_n(\mathbb{C})$) associated to π (resp. τ), and $r : {}^LG^\circ \otimes GL_n(\mathbb{C}) \to GL_{2n^2}(\mathbb{C})$ denotes the standard embedding: q is the cardinality of the residue class field of k.

Proof of Proposition A.1. We shall compute the right and left hand sides of (A.1.1) separately and then compare them. By definition of the local Langlands factor,

$$L(2s,\tau,\Lambda^2) = \frac{1}{det(I - \Lambda^2(t_\tau)q^{-2s})}.$$

On the other hand, let us recall the familiar (Poincaré polynomial) identity

$$(A.1.2)\qquad \frac{1}{det(I-AX)}=\sum_{\ell=0}^{\infty} trace\ (Sym^{\ell}(A))X^{\ell},$$

where A is an arbitrary square complex matrix, X is a sufficiently small complex variable, and $Sym^{\ell}(A)$ is the ℓ-th symmetric power of A. Applying this identity to the matrix $A=\Lambda^2(t_\tau)$, we rewrite $L(2s,\tau,\Lambda^2)$ as

$$L(2s,\tau,\Lambda)=\sum_{\ell=0}^{\infty} tr\ Sym^{\ell}(\Lambda^2(t_\tau))Y^{\ell}$$

with $Y=q^{-2s}$.

Now let us try to evaluate $\varsigma(s,W,f)$. Because all the functions which appear in the integrand defining $\varsigma(s,W,f)$ are right invariant by the standard maximal compact subgroup of H, an application of the Iwasawa decomposition to $U\setminus H$ yields the identity

$$\varsigma(s,W,f)=\int_T W_\pi(t)W_\tau^f(t,s)\delta_H^{-1}(t)dt,$$

where T denotes the maximal (split) torus

$$\left\{\begin{bmatrix} a_1 & & & & & & 0\\ & \ddots & & & & & \\ & & a_n & & & & \\ & & & 1 & & & \\ & & & & a_n^{-1} & & \\ & & & & & \ddots & \\ 0 & & & & & & a_n^{-1}\end{bmatrix} : a_i\in k^x\right\}$$

and δ_H^{-1} denotes the modulus function of the Borel subgroup of H (restricted to T). Next we need to appeal to some interesting facts about unramified Whittaker functions proved in [Casselman-Shalika] and [Kato]. Since T is also isomorphic to the maximal split torus of $M\approx GL_n$, we may regard $W_\tau(t,s)$ as $|det|^{s+\frac{n-2}{2}}$ times the restriction to T of the normalized spherical Whittaker function in $\mathcal{W}(\tau,\psi^{-1})$. Thus the general theory (cf. Prop. 6.1 of [Casselman-Shalika]) tells us right away that $W_\tau(t,s)\equiv 0$ unless $ord(a_1)\geq ord(a_2)\geq\cdots\geq ord(a_n)$, and similarly $W_\pi(t)\equiv 0$ unless $ord(a_1)\geq\cdots\geq ord(a_n)\geq 0$ (this last inequality appearing because of the simple root λ_n of G). Thus we have

$$\varsigma(s,W,f)=\sum_{\substack{\delta=(m_1,\cdots,m_n)\\ dominant}} \delta_H^{-1}(q^\delta)W_\pi(q^\delta)W_\tau(q^\delta)(q^{-(s+\frac{n-2}{2})})^{tr(\delta)}$$

where δ "dominant" means $m_1 \geq m_2 \geq \cdots \geq m_n \geq 0$, q^δ denotes the diagonal matrix $(q^{m_1}, \cdots, q^{m_n})$ embedded in H or GL_n, and $tr(\delta) = \sum_{i=1}^n m_i$.

To proceed further, we need to appeal to a deeper result about Whittaker functions, namely the explicit formula evaluating them in terms of characters of finite dimensional representations of the corresponding L-groups. Thus, for each dominant weight δ, we let χ_δ^G (resp. $\chi_\delta^{GL_n}$) denote the character of the irreducible finite dimensional holomorphic representation of $Sp_{2n}(\mathbb{C}) = {}^LG^\circ$ (resp. $GL_n(\mathbb{C})$) whose highest weight is $\delta = (m_1, \cdots, m_n)$. Then according to the main theorem of [Kato], we have

$$W_\pi(q^\delta) = \delta_G^{1/2}(q^\delta)\chi_\delta^G(t_\pi)$$

and

$$W_\tau(q^\delta) = \delta_{GL_n}^{1/2}(q^\delta)\chi_\delta^{GL_n}(t_\tau).$$

Here δ_G (resp. δ_{GL_n}) denotes the modulus function of the Borel subgroup of G (resp. GL_n) restricted to the maximal forus T. It is easy to check that $\delta_H = \delta_G^{1/2}\delta_{GL_n}^{1/2}|det|^{\frac{n-2}{2}}$. Plugging these formulas into our expression for $\varsigma(s, W, f)$, we obtain finally the identity

$$\varsigma(s, W, f) = \sum_{dominant\ \delta} \chi_\delta^G(t_\tau)\chi_\delta^{GL_n}(t_\tau)X^{tr(\delta)} = \sum_{\ell=0}\left[\sum_{\substack{\delta\ dominant \\ tr(\delta)=\ell}} \chi_\delta^G(t_\tau)\chi_\delta^{GL_n}(t_\tau)\right]X^\ell$$

with $X = q^{-s}$. Multiplying this last power series times the series for $L(2s, \tau, \Lambda^2)$, we conclude that the left hand side of (A.1.1) equals

$$(A.1.3) \qquad \sum_{\ell=0}^{\infty}\left[\sum_{2i+j=\ell} tr\ Sym^i(\Lambda^2(t_\tau))\left\{\sum_{\substack{tr(\delta)=j \\ \delta\ dominant}} \chi_\delta^G(t_\pi)\chi_\delta^{GL_n}(t_\tau)\right\}\right]X^\ell.$$

It remains to evaluate the right hand side of (A.1.1) and compare it with the expression (A.1.3). Appealing once again to the identity (A.1.2), we may rewrite this right hand side as the power series

$$\sum_{\ell=0}^{\infty} trace\ (Sym^\ell(t_\pi \otimes t_\tau))X^\ell.$$

Our task then, is to prove that $tr(Sym^\ell(t_\pi \otimes t_\tau))$ is equal to the expression in parentheses in (A.1.3). In representation theory terms, this problem may be reformulated as follows.

Let E denote $n \times 2n$ complex matrix space $\mathbb{C}^{n\times 2n}$. We may regard E as a $GL_n(\mathbb{C}) \times Sp_{2n}(\mathbb{C})$ module by letting $GL_n(\mathbb{C})$ (resp. $Sp_{2n}(\mathbb{C})$) act by left (resp. right) matrix multiplication. If $S(E^*)$ denotes the symmetric algebra of all complex-valued polynomial functions on E, then $S(E^*)$ is also a natural $GL_n(\mathbb{C}) \times Sp_{2n}(\mathbb{C})$ module, and our problem is to analyze this module.

To this end, let $I(E^*)$ denote the subalgebra of $S(E^*)$ consisting of all $G = Sp_{2n}(\mathbb{C})$ invariant polynomials. Equivalently, $I(E^*)$ is the algebra of polynomials on the space of skew-symmetric matrices $Y = XJX^t$, where $X \in E$, and $J = \begin{bmatrix} 0 & I_n \\ -I_n & 0 \end{bmatrix}$. Note $Sp_{2n}(\mathbb{C})$ acts trivially on $I(E^*)$ and $GL_n(\mathbb{C})$ acts via the antisymmetric square representation $Y \to gYg^t$; more precisely, the natural action of g in the space of homogeneous polynomials of degree i on the space of $n \times n$ antisymmetric matrices Y is equivalent to $Sym^i(\Lambda^2(g))$. Now it is known (see [Ton-That 1]) that

$$S(E^*) \approx I(E^*) \otimes H(E^*),$$

where the isomorphism is one of $GL_n(\mathbb{C}) \times Sp_{2n}(\mathbb{C})$ modules, and $H(E^*)$ is the subspace of $S(E^*)$ consisting of all G-harmonic polynomials; furthermore, if $H(E, \delta)$ denotes the subspace of $H(E^*)$ consisting of polynomials which transform under $GL_n(\mathbb{C})$ according to the irreducible representation $\rho_\delta^{GL_n}$ of $GL_n(\mathbb{C})$ of highest weight $\delta = (m_1, \cdots, m_n)$, with $m_1 \geq m_2 \geq \cdots \geq m_n \geq 0$, then as a $Sp_{2n}(\mathbb{C})$-module, $H(E, \delta)$ is equivalent to the irreducible holomorphic representation ρ_δ^G of $Sp_{2n}(\mathbb{C})$ of highest weight δ, and therefore

$$H^*(E) \approx \bigoplus_{\delta \text{ dominant}} \rho_\delta^{GL_n} \otimes \rho_\delta^G.$$

From this it follows that

$$tr\ Sym^\ell(t_\pi \otimes t_\tau) = \sum_{2i+j=\ell} \{tr\ (Sym^i(\Lambda^2(t_\tau))\} \sum_{\substack{tr(\delta)=j \\ \delta \text{ dominant}}} \chi_\delta^G(t_\pi) \chi_\delta^{GL_n}(t_\tau)$$

and so our proposition is proved.

Remark. The proof described above was initiated by the third-named author, and originally formulated by him in terms of dual reductive pairs and the oscillator representation (cf. [PSR3]). In the case of Rankin-Selberg convolutions for $GL(m) \times GL(n)$, a computation similar to the one ultimately used above was carried out in Section 2 of [JS].

(A.2) Suppose G is of type D_n. Then we must prove that

$$(A.2.1) \qquad L(2s, \tau, Sym^2)\varsigma(s, W, f) = \frac{1}{det(I - r(t_\pi \otimes t_\tau)q^{-s}}$$

Here W belongs to the Whittaker model of an unramified representation of $H = SO(V')$, and $f \in Ind\ \tau \otimes ||^{s'}$, with τ still an unramified representation of $GL_n(k)$. Thus t_π represents a conjugacy class in the group $SO_{2n}(\mathbb{C}) = {}^L H^\circ$, and t_τ is again a conjugacy class in $GL_n(\mathbb{C})$. In this case, we must appeal to [Ton-That 2] in order to analyze the $GL_n(\mathbb{C}) \times SO2n(\mathbb{C})$ module $S(E^*)$ and the right hand side of (A.2.1).

Since our local integral is now of the form (5.3.3), the computation of $\varsigma(s, W, f)$ also involves an explicit realization of the projection $i(a)$ in (5.3.3).

Finally, we mention that the case of groups of type C_n is analogous to (but simpler than) the case of type D_n. In particular, $S(E^*)$ is analyzed as a $GL_n(\mathbb{C}) \times SO_{2n+1}(\mathbb{C})$ module, with $E = \mathbb{C}^{n,2n+1}$. The integral on the left hand side of (A.2.1) is slightly different because of the appearance of the Weil representation, but the computations are surprisingly similar.

References for Appendix

[Casselman-Shalika] W. Casselman and J. Shalika, "The unramified principal series of p-adic groups II : The Whittaker function", *Compositio Math.*, Vol. 41, Fasc. 2 (1980), pp. 207 - 231.

[GPS4] S. Gelbart and I. Piatetski-Shapiro, "L-functions for $G \times GL(n)$", the preceding paper.

[JS] "On Euler Products and the Classification of Automorphic Representations I", *American J. Math.*, Vol. 103, No.3, pp. 499 - 558.

[Kato] S. Kato, "On an explicit formula for class-1 Whittaker functions on split reductive groups over p-adic fields", preprint, 1978, University of Tokyo.

[PSR3] I. Piatetski-Shapiro and S. Rallis, "Rankin Triple L-functions", preprint, Winter 1985.

[Ton-That I] T. Ton-That, "On holomorphic representations of symplectic groups", *Bull. Amer. Math. Soc.*, Vol. 81, No.6 (1975), pp. 1069-1072.

[Ton-That 2] T. Ton-That, "Lie group representations and harmonic polynomials of a matrix variable", *Trans. Amer. Math. Soc.*, Vol. 26 (1976), pp. 1 - 46.

GENERAL REFERENCES FOR PART B

[Arthur] J. Arthur, "Automorphic Representations and Number Theory", in *Canadian Mathematical Society Conference Proceedings*, Vol. 1, Providence, R.I., 1981, pp.3-51.

[Bernstein] J. Bernstein, (= I.N. Bernstein), letter to Piatetski-Shapiro, Fall 1985.

[BZ 1] I.N. Bernstein and A.V. Zelevinsky, "Induced Representations of reductive p-adic groups I", *Ann. Sci. Ecole Norm. Sup.* 4^e serie, t.10 (1977), pp. 441-472.

[BZ 2] I.N. Bernstein and A. V. Zelevinsky, "Representations of the group $GL(n, F)$ where F is a local Non-Archimedean Field", *Russian Mathematical Surveys* no.3 (1976), pp. 1-68, from Uspekhi Mat. Nauk. 31:3 (1976), 5-70.

[Borel 1] A. Borel, "Automorphic L-functions", in *Proceedings of Symposia in Pure Mathematics*, Vol. 33, Part 2, Amer. Math. Soc., Providence, R.I., 1979, pp. 27-61.

[Borel 2] A. Borel, "Linear Algebraic Groups", in *Proc. of Symposia in Pure Math.*, Vol. 9, A.M.S., Providence, R.I., 1966, pp. 3-19.

[Casselman] W. Casselman, "Introduction to the theory of admissible representations of p-adic reductive groups", mimeographed notes, 1974.

[Casselman-Shalika] W. Casselman and J. Shalika, "The unramified principal series of p-adic groups II : The Whittaker function", *Compositio Math.*, Vol. 41, Fasc. 2 (1980), pp. 207 - 231.

[Flicker] Y. Flicker, "Twisted Tensors and Euler Products", preprint, 1985-86, Harvard University.

[Garrett] P. Garrett, Rankin Triple Products attached to Holomorphic Cusp Forms, preprint, 1985, University of Minnesota.

[Gelbart] S. Gelbart, "Weil's Representation and the Spectrum of the Metaplectic Group", *Lecture Notes in Mathetmatics*, Vol. 530, Springer-Verlag, New York, 1976.

[GeJa] S. Gelbart and H. Jacquet, "A relation between automorphic representations of $GL(2)$ and $GL(3)$', *Ann. Sci. Ecole Normale Sup.*, 4^e serie, Vol. 11, (1978), pp. 471-552.

[GePS] S. Gelbart and I. Piatetski-Shapiro, "Automorphic forms and L-functions for the Unitary Group", in *Lie Group Representations II, Lecture Notes in Mathematics*, Vol. 1041, Springer-Verlag, New York, 1984, pp. 141-184.

[GePS2] S. Gelbart and I. Piatetski-Shapiro, "Distinguished representations and modular forms of half-integral weight", *Inventiones Mathematicae*, 59 (1980), pp. 145-188.

[GePS3] S. Gelbart and I. Piatetski-Shapiro, "On Shimura's correspondence for modular forms of half-integral weight", *Automorphic Forms, Representation theory and Arithmetic* (Bombay, 1979), Tata Institute of Fund. Research Studies in Math., No. 10, Bombay, 1981, pp. 1-39.

[Gel Kaz] I.M. Gelfand and D. Kazhdan, "Representations of the group $GL(n, K)$ where K is a local field", in *Lie Groups and their Representations*, Halsted, New York, 1975, pp. 95 - 118.

[Godement-Jacquet] R. Godement and H. Jacquet, "Zeta-functions of Simple Algebras", *Lecture Notes in Mathematics*, Vol. 260, Springer-Verlag, New York, 1972.

[Howe] R. Howe, "Classification of Irreducible Representations of $GL_2(F)$", preprint, I.H.E.S., Bures-sur-Yvette, France, 1978.

[HoPS] R. Howe and I. Piatetski-Shapiro, "Some examples of automorphic forms on Sp_4", *Duke Math. Journal*, Vol. 50, No. 1, (1983), pp. 55-106.

[Jacquet] H. Jacquet, "Automorphic Forms on $GL(2)$, II", *Lecture Notes in Mathematics*, Vol. 278, Springer-Verlag, New York, 1972.

[JPSS1] H. Jacquet, I. Piatetski-Shapiro and J. Shalika, "Rankin-Selberg Convolutions", *Amer. J. Math.*, Vol. 105, (1983), pp. 367-464.

[JPSS2] H. Jacquet, I. Piatetski-Shapiro and J. Shalika, "Automorphic Forms on $GL(3)$ II", *Annals of Math.*, 109 (1979), pp. 213-258.

[JS] H. Jacquet and J. Shalika, "On Euler Products and the Classification of Automorphic Representations I", *Amer. J. Math.*, Vol. 103, No.3, pp. 499-558.

[Jacobson] N. Jacobson, *Basic Algebra I*, W.H. Freeman and Company, San Francisco, 1974.

[Kato] S. Kato, "On the explicit formula for class-1 Whittaker functions on split reductive groups over p-adic fields", preprint 1978, University of Tokyo.

[Kaz Pat] D. Kazhdan and S.J. Patterson, "Metaplectic Forms", *Pub. Institut des Hautes Etudes scient.*, no. 59, 1984.

[Kubota] T. Kubota, "Automorphic functions and the reciprocity law in a number field", mimeographed notes, Kyoto University, 1969.

[Langlands 1] R.P. Langlands, "Problems in the Theory of Automorphic Forms", in *Lecture Notes in Mathematics*, Vol. 170, Springer-Verlag, New York, 1970, pp. 18-86.

[Langlands 2] R.P. Langlands, *Euler Products*, Yale University Press, James K. Whitmore Lectures, 1967.

[Langlands 3] R.P. Langlands, "On the Functional Equations satisfied by Eisenstein Series', *Lecture Notes in Mathematics*, Vol. 544, Springer-Verlag, New York, 1976.

[Moore] C. Moore, "Group extensions of p-adic linear groups", *Publ. Math. I.H.E.S.* No.35 (1968).

[PS] I. Piatetski-Shapiro, "Euler Subgroups", in *Lie Groups and their Representations*, Halsted, New York, 1975.

[PS2] I. Piatetski-Shapiro, "Tate Theory for Reductive Groups and Distinguished Representations", *Proceedings of the International Congress of Math.*, Helsinki, 1978.

[PSR1] I. Piatetski-Shapiro and S. Rallis, "L-functions for the Classical Groups", Part I of this volume; see also the announcement "L-functions of Automorphic Forms on Simple Classical groups" Chapter 10 in *Modular Forms*, ed. R. Rankin, Ellis Horward, 1984, pp. 251 - 262.

[PSR2] I. Piatetski-Shapiro and S. Rallis, " ε-factors of Representations of Classical Groups", *Proceedings of the National Academy of Sciences*, U.S.A. ; Vol. 83, pp.4589 - 4593, July 1986.

[PSR3] I. Piatetski-Shapiro and S. Rallis, Rankin Triple L-funtions, preprint, Winter 1985.

[PSS1] I. Piatetski-Shapiro and D. Soudry, "L and ε funtions for $GSp(4) \times GL(2)$, *Proc. Nat. Acad. Sci.*, U.S.A., Vol. 81, June 1984, pp. 3924 - 3927.

[PSS2] I. Piatetski-Shapiro and D. Soudry, "Automorphic Forms on the Symplectic Group of Order four", preprint, Institut des Hautes Etudes Scientifiques, Bures-sur-Yvette, July 1983.

[Rankin] R. Rankin, Contributions to the theory of Ramanujan's function $\tau(n)$ and similar arithmetical functions, I and II", Proc. Camb. Phil. Soc. 35, (1939), pp. 351-356 and 357-372.

[Selberg] A. Selberg, "Bemerkungen uber eine Dirichletsche Reihe, die mit der Theorie der Modulformer nahe verbunden ist", Arch. Math. Naturvid. 43 (1940), pp. 47 - 50.

[Shahidi] F. Shahidi, "On Certain L-functions", *Amer. J. Math.*, Vol. 103, No.2, 1981, pp. 297-355.

[Shahidi 2] F. Shahidi, On the Ramanujan Conjecture and the finiteness of poles of certain L-functions, preprint, 1986.

[Shalika] J. Shalika, "The multiplicity one theorem for GL_n", *Annals of Math.* Vol. 100, No.1, (1974), pp. 171 - 193.

[Shimura] G. Shimura, "On the holomorphy of certain Dirichlet series", *Proc. London Math. Soc.* (3) 31 (1975), pp. 79 - 98.

[Shimura] G. Shimura, "On modular forms of half-integral weight", *Annals of Math.*, Vol. 97 (1973), pp. 440 - 481.

[Soudry] D. Soudry, "The L and γ factors for generic representations of $GSp(4,k) \times GL(2,k)$ over a local nonarchimedean field k", *Duke Math. J.*, Vol. 51, No.2, June 1984, pp. 355 - 394.

[Ton-That 1] T. Ton-That, "On holomorphic representations of symplectic groups", *Bull. Amer. Math. Soc.* Vol. 81, No. 6 (1975), pp. 1069 - 1072.

[Ton-That 2] T. Ton-That, "Lie group representations and harmonic polynomials of a matrix variable", *Trans. Amer. Math. Soc.*, Vol. 126 (1976), pp. 1 - 46.

[Weil] A. Weil, "Sur certaines groupes d'opérateurs unitaires", *Acta Math.* 111 (1964), pp. 143 - 211.

General Index

Index for Notation

Vol. 1090: Differential Geometry of Submanifolds. Proceedings, 1984. Edited by K. Kenmotsu. VI, 132 pages. 1984.

Vol. 1091: Multifunctions and Integrands. Proceedings, 1983. Edited by G. Salinetti. V, 234 pages. 1984.

Vol. 1092: Complete Intersections. Seminar, 1983. Edited by S. Greco and R. Strano. VII, 299 pages. 1984.

Vol. 1093: A. Prestel, Lectures on Formally Real Fields. XI, 125 pages. 1984.

Vol. 1094: Analyse Complexe. Proceedings, 1983. Edité par E. Amar, R. Gay et Nguyen Thanh Van. IX, 184 pages. 1984.

Vol. 1095: Stochastic Analysis and Applications. Proceedings, 1983. Edited by A. Truman and D. Williams. V, 199 pages. 1984.

Vol. 1096: Théorie du Potentiel. Proceedings, 1983. Edité par G. Mokobodzki et D. Pinchon. IX, 601 pages. 1984.

Vol. 1097: R.M. Dudley, H. Kunita, F. Ledrappier, École d'Éte de Probabilités de Saint-Flour XII – 1982. Edité par P.L. Hennequin. X, 396 pages. 1984.

Vol. 1098: Groups – Korea 1983. Proceedings. Edited by A.C. Kim and B.H. Neumann. VII, 183 pages. 1984.

Vol. 1099: C.M. Ringel, Tame Algebras and Integral Quadratic Forms. XIII, 376 pages. 1984.

Vol. 1100: V. Ivrii, Precise Spectral Asymptotics for Elliptic Operators Acting in Fiberings over Manifolds with Boundary. V, 237 pages. 1984.

Vol. 1101: V. Cossart, J. Giraud, U. Orbanz, Resolution of Surface Singularities. Seminar. VII, 132 pages. 1984.

Vol. 1102: A. Verona, Stratified Mappings – Structure and Triangulability. IX, 160 pages. 1984.

Vol. 1103: Models and Sets. Proceedings, Logic Colloquium, 1983, Part I. Edited by G.H. Müller and M.M. Richter. VIII, 484 pages. 1984.

Vol. 1104: Computation and Proof Theory. Proceedings, Logic Colloquium, 1983, Part II. Edited by M.M. Richter, E. Börger, W. Oberschelp, B. Schinzel and W. Thomas. VIII, 475 pages. 1984.

Vol. 1105: Rational Approximation and Interpolation. Proceedings, 1983. Edited by P.R. Graves-Morris, E.B. Saff and R.S. Varga. XII, 528 pages. 1984.

Vol. 1106: C.T. Chong, Techniques of Admissible Recursion Theory. IX, 214 pages. 1984.

Vol. 1107: Nonlinear Analysis and Optimization. Proceedings, 1982. Edited by C. Vinti. V, 224 pages. 1984.

Vol. 1108: Global Analysis – Studies and Applications I. Edited by Yu.G. Borisovich and Yu.E. Gliklikh. V, 301 pages. 1984.

Vol. 1109: Stochastic Aspects of Classical and Quantum Systems. Proceedings, 1983. Edited by S. Albeverio, P. Combe and M. Sirugue-Collin. IX, 227 pages. 1985.

Vol. 1110: R. Jajte, Strong Limit Theorems in Non-Commutative Probability. VI, 152 pages. 1985.

Vol. 1111: Arbeitstagung Bonn 1984. Proceedings. Edited by F. Hirzebruch, J. Schwermer and S. Suter. V, 481 pages. 1985.

Vol. 1112: Products of Conjugacy Classes in Groups. Edited by Z. Arad and M. Herzog. V, 244 pages. 1985.

Vol. 1113: P. Antosik, C. Swartz, Matrix Methods in Analysis. IV, 114 pages. 1985.

Vol. 1114: Zahlentheoretische Analysis. Seminar. Herausgegeben von E. Hlawka. V, 157 Seiten. 1985.

Vol. 1115: J. Moulin Ollagnier, Ergodic Theory and Statistical Mechanics. VI, 147 pages. 1985.

Vol. 1116: S. Stolz, Hochzusammenhängende Mannigfaltigkeiten und ihre Ränder. XXIII, 134 Seiten. 1985.

Vol. 1117: D.J. Aldous, J.A. Ibragimov, J. Jacod, Ecole d'Été de Probabilités de Saint-Flour XIII – 1983. Édité par P.L. Hennequin. IX, 409 pages. 1985.

Vol. 1118: Grossissements de filtrations: exemples et applications. Seminaire, 1982/83. Edité par Th. Jeulin et M. Yor. V, 315 pages. 1985.

Vol. 1119: Recent Mathematical Methods in Dynamic Programming. Proceedings, 1984. Edited by I. Capuzzo Dolcetta, W.H. Fleming and T. Zolezzi. VI, 202 pages. 1985.

Vol. 1120: K. Jarosz, Perturbations of Banach Algebras. V, 118 pages. 1985.

Vol. 1121: Singularities and Constructive Methods for Their Treatment. Proceedings, 1983. Edited by P. Grisvard, W. Wendland and J.R. Whiteman. IX, 346 pages. 1985.

Vol. 1122: Number Theory. Proceedings, 1984. Edited by K. Alladi. VII, 217 pages. 1985.

Vol. 1123: Séminaire de Probabilités XIX 1983/84. Proceedings. Edité par J. Azéma et M. Yor. IV, 504 pages. 1985.

Vol. 1124: Algebraic Geometry, Sitges (Barcelona) 1983. Proceedings. Edited by E. Casas-Alvero, G.E. Welters and S. Xambó-Descamps. XI, 416 pages. 1985.

Vol. 1125: Dynamical Systems and Bifurcations. Proceedings, 1984. Edited by B.L.J. Braaksma, H.W. Broer and F. Takens. V, 129 pages. 1985.

Vol. 1126: Algebraic and Geometric Topology. Proceedings, 1983. Edited by A. Ranicki, N. Levitt and F. Quinn. V, 423 pages. 1985.

Vol. 1127: Numerical Methods in Fluid Dynamics. Seminar. Edited by F. Brezzi, VII, 333 pages. 1985.

Vol. 1128: J. Elschner, Singular Ordinary Differential Operators and Pseudodifferential Equations. 200 pages. 1985.

Vol. 1129: Numerical Analysis, Lancaster 1984. Proceedings. Edited by P.R. Turner. XIV, 179 pages. 1985.

Vol. 1130: Methods in Mathematical Logic. Proceedings, 1983. Edited by C.A. Di Prisco. VII, 407 pages. 1985.

Vol. 1131: K. Sundaresan, S. Swaminathan, Geometry and Nonlinear Analysis in Banach Spaces. III, 116 pages. 1985.

Vol. 1132: Operator Algebras and their Connections with Topology and Ergodic Theory. Proceedings, 1983. Edited by H. Araki, C.C. Moore, Ş. Strătilă and C. Voiculescu. VI, 594 pages. 1985.

Vol. 1133: K.C. Kiwiel, Methods of Descent for Nondifferentiable Optimization. VI, 362 pages. 1985.

Vol. 1134: G.P. Galdi, S. Rionero, Weighted Energy Methods in Fluid Dynamics and Elasticity. VII, 126 pages. 1985.

Vol. 1135: Number Theory, New York 1983–84. Seminar. Edited by D.V. Chudnovsky, G.V. Chudnovsky, H. Cohn and M.B. Nathanson. V, 283 pages. 1985.

Vol. 1136: Quantum Probability and Applications II. Proceedings, 1984. Edited by L. Accardi and W. von Waldenfels. VI, 534 pages. 1985.

Vol. 1137: Xiao G., Surfaces fibrées en courbes de genre deux. IX, 103 pages. 1985.

Vol. 1138: A. Ocneanu, Actions of Discrete Amenable Groups on von Neumann Algebras. V, 115 pages. 1985.

Vol. 1139: Differential Geometric Methods in Mathematical Physics. Proceedings, 1983. Edited by H. D. Doebner and J. D. Hennig. VI, 337 pages. 1985.

Vol. 1140: S. Donkin, Rational Representations of Algebraic Groups. VII, 254 pages. 1985.

Vol. 1141: Recursion Theory Week. Proceedings, 1984. Edited by H.-D. Ebbinghaus, G.H. Müller and G.E. Sacks. IX, 418 pages. 1985.

Vol. 1142: Orders and their Applications. Proceedings, 1984. Edited by I. Reiner and K. W. Roggenkamp. X, 306 pages. 1985.

Vol. 1143: A. Krieg, Modular Forms on Half-Spaces of Quaternions. XIII, 203 pages. 1985.

Vol. 1144: Knot Theory and Manifolds. Proceedings, 1983. Edited by D. Rolfsen. V, 163 pages. 1985.

Vol. 1145: G. Winkler, Choquet Order and Simplices. VI, 143 pages. 1985.

Vol. 1146: Séminaire d'Algèbre Paul Dubreil et Marie-Paule Malliavin. Proceedings, 1983–1984. Edité par M.-P. Malliavin. IV, 420 pages. 1985.

Vol. 1147: M. Wschebor, Surfaces Aléatoires. VII, 111 pages. 1985.

Vol. 1148: Mark A. Kon, Probability Distributions in Quantum Statistical Mechanics. V, 121 pages. 1985.

Vol. 1149: Universal Algebra and Lattice Theory. Proceedings, 1984. Edited by S. D. Comer. VI, 282 pages. 1985.

Vol. 1150: B. Kawohl, Rearrangements and Convexity of Level Sets in PDE. V, 136 pages. 1985.

Vol 1151: Ordinary and Partial Differential Equations. Proceedings, 1984. Edited by B.D. Sleeman and R.J. Jarvis. XIV, 357 pages. 1985.

Vol. 1152: H. Widom, Asymptotic Expansions for Pseudodifferential Operators on Bounded Domains. V, 150 pages. 1985.

Vol. 1153: Probability in Banach Spaces V. Proceedings, 1984. Edited by A. Beck, R. Dudley, M. Hahn, J. Kuelbs and M. Marcus. VI, 457 pages. 1985.

Vol. 1154: D.S. Naidu, A.K. Rao, Singular Pertubation Analysis of Discrete Control Systems. IX, 195 pages. 1985.

Vol. 1155: Stability Problems for Stochastic Models. Proceedings, 1984. Edited by V.V. Kalashnikov and V.M. Zolotarev. VI, 447 pages. 1985.

Vol. 1156: Global Differential Geometry and Global Analysis 1984. Proceedings, 1984. Edited by D. Ferus, R.B. Gardner, S. Helgason and U. Simon. V, 339 pages. 1985.

Vol. 1157: H. Levine, Classifying Immersions into $\mathbb{R}^4$ over Stable Maps of 3-Manifolds into $\mathbb{R}^2$. V, 163 pages. 1985.

Vol. 1158: Stochastic Processes – Mathematics and Physics. Proceedings, 1984. Edited by S. Albeverio, Ph. Blanchard and L. Streit. VI, 230 pages. 1986.

Vol. 1159: Schrödinger Operators, Como 1984. Seminar. Edited by S. Graffi. VIII, 272 pages. 1986.

Vol. 1160: J.-C. van der Meer, The Hamiltonian Hopf Bifurcation. VI, 115 pages. 1985.

Vol. 1161: Harmonic Mappings and Minimal Immersions, Montecatini 1984. Seminar. Edited by E. Giusti. VII, 285 pages. 1985.

Vol. 1162: S.J.L. van Eijndhoven, J. de Graaf, Trajectory Spaces, Generalized Functions and Unbounded Operators. IV, 272 pages. 1985.

Vol. 1163: Iteration Theory and its Functional Equations. Proceedings, 1984. Edited by R. Liedl, L. Reich and Gy. Targonski. VIII, 231 pages. 1985.

Vol. 1164: M. Meschiari, J.H. Rawnsley, S. Salamon, Geometry Seminar "Luigi Bianchi" II – 1984. Edited by E. Vesentini. VI, 224 pages. 1985.

Vol. 1165: Seminar on Deformations. Proceedings, 1982/84. Edited by J. Ławrynowicz. IX, 331 pages. 1985.

Vol. 1166: Banach Spaces. Proceedings, 1984. Edited by N. Kalton and E. Saab. VI, 199 pages. 1985.

Vol. 1167: Geometry and Topology. Proceedings, 1983–84. Edited by J. Alexander and J. Harer. VI, 292 pages. 1985.

Vol. 1168: S.S. Agaian, Hadamard Matrices and their Applications. III, 227 pages. 1985.

Vol. 1169: W.A. Light, E.W. Cheney, Approximation Theory in Tensor Product Spaces. VII, 157 pages. 1985.

Vol. 1170: B.S. Thomson, Real Functions. VII, 229 pages. 1985.

Vol. 1171: Polynômes Orthogonaux et Applications. Proceedings, 1984. Edité par C. Brezinski, A. Draux, A.P. Magnus, P. Maroni et A. Ronveaux. XXXVII, 584 pages. 1985.

Vol. 1172: Algebraic Topology, Göttingen 1984. Proceedings. Edited by L. Smith. VI, 209 pages. 1985.

Vol. 1173: H. Delfs, M. Knebusch, Locally Semialgebraic Spaces. XVI, 329 pages. 1985.

Vol. 1174: Categories in Continuum Physics, Buffalo 1982. Seminar. Edited by F.W. Lawvere and S.H. Schanuel. V, 126 pages. 1986.

Vol. 1175: K. Mathiak, Valuations of Skew Fields and Projective Hjelmslev Spaces. VII, 116pages. 1986.

Vol. 1176: R.R. Bruner, J.P. May, J.E. McClure, M. Steinberger, H_∞ Ring Spectra and their Applications. VII, 388 pages. 1986.

Vol. 1177: Representation Theory I. Finite Dimensional Algebras. Proceedings, 1984. Edited by V. Dlab, P. Gabriel and G. Michler. XV, 340 pages. 1986.

Vol. 1178: Representation Theory II. Groups and Orders. Proceedings, 1984. Edited by V. Dlab, P. Gabriel and G. Michler. XV, 370 pages. 1986.

Vol. 1179: Shi J.-Y. The Kazhdan-Lusztig Cells in Certain Affine Weyl Groups. X, 307 pages. 1986.

Vol. 1180: R. Carmona, H. Kesten, J.B. Walsh, École d'Été de Probabilités de Saint-Flour XIV – 1984. Édité par P.L. Hennequin. X, 438 pages. 1986.

Vol. 1181: Buildings and the Geometry of Diagrams, Como 1984. Seminar. Edited by L. Rosati. VII, 277 pages. 1986.

Vol. 1182: S. Shelah, Around Classification Theory of Models. VII, 279 pages. 1986.

Vol. 1183: Algebra, Algebraic Topology and their Interactions. Proceedings, 1983. Edited by J.-E. Roos. XI, 396 pages. 1986.

Vol. 1184: W. Arendt, A. Grabosch, G. Greiner, U. Groh, H.P. Lotz, U. Moustakas, R. Nagel, F. Neubrander, U. Schlotterbeck, One-parameter Semigroups of Positive Operators. Edited by R. Nagel. X, 460 pages. 1986.

Vol. 1185: Group Theory, Beijing 1984. Proceedings. Edited by Tuan H.F. V, 403 pages. 1986.

Vol. 1186: Lyapunov Exponents. Proceedings, 1984. Edited by L. Arnold and V. Wihstutz. VI, 374 pages. 1986.

Vol. 1187: Y. Diers, Categories of Boolean Sheaves of Simple Algebras. VI, 168 pages. 1986.

Vol. 1188: Fonctions de Plusieurs Variables Complexes V. Séminaire, 1979–85. Edité par François Norguet. VI, 306 pages. 1986.

Vol. 1189: J. Lukeš, J. Malý, L. Zajíček, Fine Topology Methods in Real Analysis and Potential Theory. X, 472 pages. 1986.

Vol. 1190: Optimization and Related Fields. Proceedings, 1984. Edited by R. Conti, E. De Giorgi and F. Giannessi. VIII, 419 pages. 1986.

Vol. 1191: A.R. Its, V.Yu. Novokshenov, The Isomonodromic Deformation Method in the Theory of Painlevé Equations. IV, 313 pages. 1986.

Vol. 1192: Equadiff 6. Proceedings, 1985. Edited by J. Vosmansky and M. Zlámal. XXIII, 404 pages. 1986.

Vol. 1193: Geometrical and Statistical Aspects of Probability in Banach Spaces. Proceedings, 1985. Edited by X. Fernique, B. Heinkel, M.B. Marcus and P.A. Meyer. IV, 128 pages. 1986.

Vol. 1194: Complex Analysis and Algebraic Geometry. Proceedings, 1985. Edited by H. Grauert. VI, 235 pages. 1986.

Vol.1195: J.M. Barbosa, A.G. Colares, Minimal Surfaces in $\mathbb{R}^3$. X, 124 pages. 1986.

Vol. 1196: E. Casas-Alvero, S. Xambó-Descamps, The Enumerative Theory of Conics after Halphen. IX, 130 pages. 1986.

Vol. 1197: Ring Theory. Proceedings, 1985. Edited by F.M.J. van Oystaeyen. V, 231 pages. 1986.

Vol. 1198: Séminaire d'Analyse, P. Lelong – P. Dolbeault – H. Skoda. Seminar 1983/84. X, 260 pages. 1986.

Vol. 1199: Analytic Theory of Continued Fractions II. Proceedings, 1985. Edited by W.J. Thron. VI, 299 pages. 1986.

Vol. 1200: V.D. Milman, G. Schechtman, Asymptotic Theory of Finite Dimensional Normed Spaces. With an Appendix by M. Gromov. VIII, 156 pages. 1986.